Safety, Health, *and* Environment *for Process Technicians*

Safety, Health, *and* Environment *for Process Technicians*

by

Charles E. Thomas

UHAI

Uhai Publishing, Inc.
1731 Helderberg Trail
Berne, NY 12023
Toll free 1-877-842-4782

NOTICE TO THE READER

Publisher does not warrant or guarantee any of the products described herein or perform any independent analysis in connection with any of the product information contained herein. Publisher does not assume, and expressly disclaims, any obligation to obtain and include information other than that provided to it by the manufacturer.

The reader is expressly warned to consider and adopt all safety precautions that might be indicated by the activities herein and to avoid all potential hazards. By following the instructions contained herein, the reader willingly assumes all risks in connections with such instructions.

The publisher makes no representation or warranties of any kind, including but not limited to, the warranties of fitness for particular purpose or merchantability, nor are any such representations implied with respect to the material set forth herein, and the publisher takes no responsibility with respect to such material. The publisher shall not be liable for any special, consequential, or exemplary damages resulting, in whole or part, from the readers' use of, or reliance upon, this material.

Cover Design:
Bi-County Graphics

Cover Photograph:
PhotoDisc

Photographs courtesy of:
Moore memorial Public Library and Charles E. Thomas

Illustrations by:
C. Thomas/ Bi-County Graphics

Printed in the United States of America

For more information, contact:
Uhai Publishing, Inc.
1731 Helderberg Trail
Berne, NY 12023.
Toll free 1-877-842-4782

2 3 4 5 6 7 8 9 10 XXX 03 02 01

Preface

Over the last two decades, technology has advanced faster than our ability to disseminate it into our existing workforce. The potential for catastrophic events is greater today than at any other time in our history. Areas of specialization such as process operations require new approaches to safety program development and management. Most safety texts are designed to focus on the broad occupation of the safety professional and manager. This text is written for the process technician who operates and maintains chemical plants and refineries, food processing plants, paper and pulp plants, or utilities.

Safety, Health, and Environment for Process Technicians covers government mandated training and foundational aspects of apprentice technician safety training. Essential topics discussed in this text include:

- Hazard Communication 29 CFR 1910.1200
- HAZWOPER 29 CFR 1910.120
- Respiratory Protection 29 CFR 1910.134
- Permit Systems- Hot Work, Confined Space Entry
- The Control of Hazardous Energy (Lockout/Tagout) CFR 1910.147
- Process Safety Management 29 CFR 1910.119
- Personal Protective Equipment 29 CFR 1910.133
- Fire Prevention, Protection and Control 29 CFR 1910.157
- Department of Transportation (DOT) 29 CFR 171-177
- Environmental Standards- RCRA, CERCLA, TSCA, and so on
- Occupational Noise Exposure 29 CFR 1910.95

An *Instructor's Guide* is available to accompany *Safety, Health, and Environment for Process Technicians*. It includes answers to end-of-chapter questions, a test for each chapter, and transparency masters.

Acknowledgments

Safety, Health, and Environment for Process Technicians is the product of many hours of research and work. The author would like to thank Steve Ames of EXXON Baytown Complex, Ray Clegg of Chevron, Bruce Richardson of DuPont, Tom Derryberry, Wiley Carter, Jerry Brown, Tom Hebert of Bayer Corporation, and Rick Guzman of Celanese.

The author and the publisher would like to thank Dan Crowl from Michigan Technological University and Steven Stokely from Lamar University for reading the manuscript and providing suggestions for making this the best possible textbook on occupational safety and health for process plant operators.

Your comments would be appreciated. If you have ideas for improving this textbook, please send them to us in care of Uhai Publishing, Inc.

Charles E. Thomas M.S.

Table Of Contents

Chapter 1

Working in the Chemical Processing Industry

OBJECTIVES

- *Describe the chemical processing industry.*
- *Identify the attitudes and behaviors of safe and environmentally sound work habits.*
- *Contrast safety, health, and environmental issues with the performance of job tasks.*
- *Classify the safety roles and responsibilities of process technicians.*
- *Contrast the explosion of the SS Grandcamp in Texas City with the 1984 vapor release at the Union Carbide India Limited of Bhopal, India.*
- *Describe the impact of the Phillips, Houston, Texas vapor release and explosion in 1989 and the ARCO, Houston, Texas explosion in 1990 on the development of the process safety management standard.*

KEY TERMS

American Chemical Society (ACS)—a self-governed individual membership organization with 155,000 members - 60% from industry.

Chemical processing industry (CPI)—broad term used to describe chemical plants and refineries, power plants, food processing, paper, pulp, and city utilities.

Community Action Emergency Response (CAER)—programs designed to respond to industrial emergencies.

Process technology— the study and application of the scientific principles associated with the operation and maintenance of the chemical processing industry.

1.2 THE PROCESS TECHNICIAN AND THE CHEMICAL PROCESSING INDUSTRY

It is easy to describe the roles of a doctor, a lawyer and a school teacher because we have grown up watching them on television, hearing our friends and parents describe them, and interacting with them ourselves. We have a good idea which occupations make the most money and the education credentials required to practice the professions. But since the role of the process technician or operator is such a specialized occupation, geographically and occupationally, it is a little more difficult to describe.

Process operators are typically found in areas that are heavily industrialized. The silicon valley for process operations appears to be the gulf coast with a focal point near Houston. There are two or three other regional areas in the United States where large numbers of process technicians can be found.

The roles of process operators are best defined by the chemical or process they are working with. There are hundreds of different process arrangements for refinery operations, petrochemical operations, and gas process operations. Process technicians work in chemical plants and refineries, the pulp and paper industry, food processing plants, power plants, city utilities, and many other areas. The roles and responsibilities of process technicians will vary depending on the areas they are working in. However, common relationships across the occupation include:

- **Equipment and technology**: valves, piping, pumps, compressors, tanks, steam turbines, motors, heat exchangers, cooling towers, boilers, furnaces, reactors, distillation columns, flare systems, waste water treatment systems, lubrication systems, refrigeration systems, plastics plant systems, auxiliary systems, hydraulic systems, specialized equipment, symbols and diagrams, and so on.

- **Instrument and electrical**: instrumentation, control loops, computers, electrical systems, interlocks, and permissives.
- **Maintenance**: using hand tools, performing minor maintenance, coordinating equipment maintenance activities, filling oil reservoirs, and housekeeping.
- **Operation**: start and stop processes, controlling industrial processes with computer automation, catching samples, making rounds, filling out check lists, filling out log books, and troubleshooting.
- **Math and science**: basic math, statistics, energy, heat, heat transfer, pressure, temperature, fluid flow, level, applied chemistry, applied physics, statistical process control, control charts, and so on.
- **Language**: reading, writing, and communication.
- **Computer technology**: word processing, spread sheets, data bases, and Internet.
- **Safety, health, and environmental**: general safety rules, process safety management, hazard communication, HAZWOPER, respiratory protection, PPE, fire protection, department of transportation, environmental regulations, permits, and so on.
- **Human relations**: interpersonal skills, working in self directed work teams, attitudes and behaviors, listening skills, communication, motivation, goal setting, and time management.
- **College skills**: well-rounded, broad-based education. Future technicians will need to have a variety of educational experiences in order to address the needs of a rapidly changing workplace.

According to the American Chemical Society, there are over 500,000 process technicians in the United States. Process operators fall into an educational category referred to as process technology. Process technology is defined as the study and application of the scientific principles associated with the operation and maintenance of the chemical processing industry.

Records indicate that process technicians are safer at their worksites than at their own homes. The chemical processing industry (CPI) has established a number of safety systems to protect their employees, facilities, and communities. These safety systems will be discussed during the course of this text.

Attitudes and Behaviors

Most employers have a screening device in their pre-employment interviews that check for specific safety attitudes and behaviors. Strong candidates for employment who demonstrate poor attitudes and behaviors toward safety will find it difficult to pass initial interviews. Typically, the employer will ask the candidate this type of question: "If you were given a direct order by your immediate supervisor to perform a task that you felt was unsafe, how would you respond?" The answer to this type of question will determine whether the candidate will be hired by this company. The only correct response is, "I do not perform unsafe acts and would not be able to comply with my supervisor's request." If the interviewer then asked, "So you would be insubordinat?" The response again should be, "I do not perform unsafe acts."

Industrial employers are also concerned about environmental issues and a prospective employee's attitude toward the environment. A typical interview question might be, " You have been assigned to the environmental control section of our plant that treats waste water before releasing it into the river. Three days after your initial assignment, a heavy storm moves over the area flooding neighboring communities and overwhelming the plant's waste treatment system. The plant's sewer system is flooded, the biological system is saturated, primary and secondary clarifiers are at full capacity, and the lagoons are almost full. Your supervisor calls you in for a private meeting to discuss options including opening the flood gates to release a small amount of the untreated effluent before the dikes break, and holding up all material to treat the situation according to environmental regulations. Which option would you choose and why?

Environmental issues are typically associated with fines and prison sentences. Informed technicians know how to respond to this scenario. The flood gates should not be opened under penalty of law. Interviewers could ask questions about littering, disposing of motor oil, new environmental regulations, and housekeeping. An applicant's answers to these types of questions will give an industrial employer insight into his or her environmental values.

The attitudes and behaviors for safe and environmentally sound work habits include:

- complying with environmental rules and regulations
- responding to environmental spills and releases
- reporting unsafe acts and conditions
- maintaining good housekeeping
- obeying safety rules
- studying safety and environmental standards
- following standard operating procedures
- performing minor maintenance on leaking equipment

During routine operation a process technician will encounter safety, health and environmental issues. Leaking pumps, pipes, and drums create environmental and safety problems. These spills could be in reportable or non-reportable quantities and must be addressed immediately by the technician. Poor housekeeping has been associated with increased safety incidents. Putting off reporting of unsafe acts and conditions directly impacts a plant's safety record. It is important for a process technician to obey all safety rules and follow their company's operational procedures.

New Technology and Potential for Catastrophic Events

Over the past twenty years, a number of advances have taken place that have revolutionized the chemical processing industry. The number one advance has been in the area of automation. The CPI can now produce more with less people because of new technology. Several companies that employed over 12,000 employees during the 1980s are now producing two to three times the product with only 2,000 employees. However, two of these companies were sold because of environmental issues and then were modernized, resold, rightsized, and automated.

As equipment and technology improved, older workers made way for the new generation. Job descriptions for process technicians were torn up and rewritten. It quickly became clear to the chemical processing industry that a highly educated and adaptable workforce was needed. Industrial manufacturers turned to their local community colleges and universities. Initial programs developed in 1992 were crude but by 1998 a standardized state curriculum was adopted in Texas and Louisiana. Several other large state efforts are underway and should see major steps toward standardization by the turn of the century. The development of this curriculum had its roots in an industrial partnership initiated by educational alliances, the American Chemical Society, and industry. (It should be emphasized that this was a national effort rather than a single state's effort.)

Along with this highly technical workforce comes a number of risks. In order to describe these risks we will need to look at this problem in two parts: the make-up of the workforce and the technology boom. Since the new workforce is a mix of the older "baby boomer" generation and a sprinkling of recent graduates of process technology programs, an unbalanced arrangement exists. Over the next five to ten years huge numbers of this workforce will retire, leaving a void that will be filled by our local community colleges and universities. During this transition, advances will continue in technology. New processes will be developed and technology will quickly fill the void with the latest and greatest. Global competition is so intense that many companies will be forced to apply cutting edge technology in order to stay in business. Most of this technology will be integrated with existing plant systems and applied for the first time by this mixed workforce. Because of increased capacity and productivity, it will be possible for a technician to push a button and cause a major incident.

The American Chemical Society projects that the present responsibilities of process technicians, chemists, and engineers will begin to merge. Another concept that is beginning to gain in popularity is the merging of process, electrical, instrument, and mechanical technicians. The process technician of the future will have a degree and will work with advanced electronic systems and equipment. Process technology programs presently exist in Texas, Louisiana, Illinois, Michigan, Utah, California, Delaware, South Carolina, North Dakota, Ohio, Alabama, Kentucky, and Colorado.

1.3 TEXAS CITY EXPLOSION

On the cool spring morning of April 16, 1947, shortly after 8 a.m., Texas City residents experienced one of the worst industrial accidents to occur in the United States. The SS Grandcamp, a French ship, was anchored in Galveston Bay beside the new Monsanto Chemical Company plant. Around 8:00 a.m., the crew noticed that a small fire probably ignited by a carelessly discarded cigarette had started in the number four hold. Inside the hold, longshoremen had been loading a crystalline powder, namely ammonium nitrate, which was used as a fertilizer for crops or as an additive for TNT. (Ammonium nitrate has about half the explosive capacity as TNT but many of the hazards associated with this chemical were not generally known.)

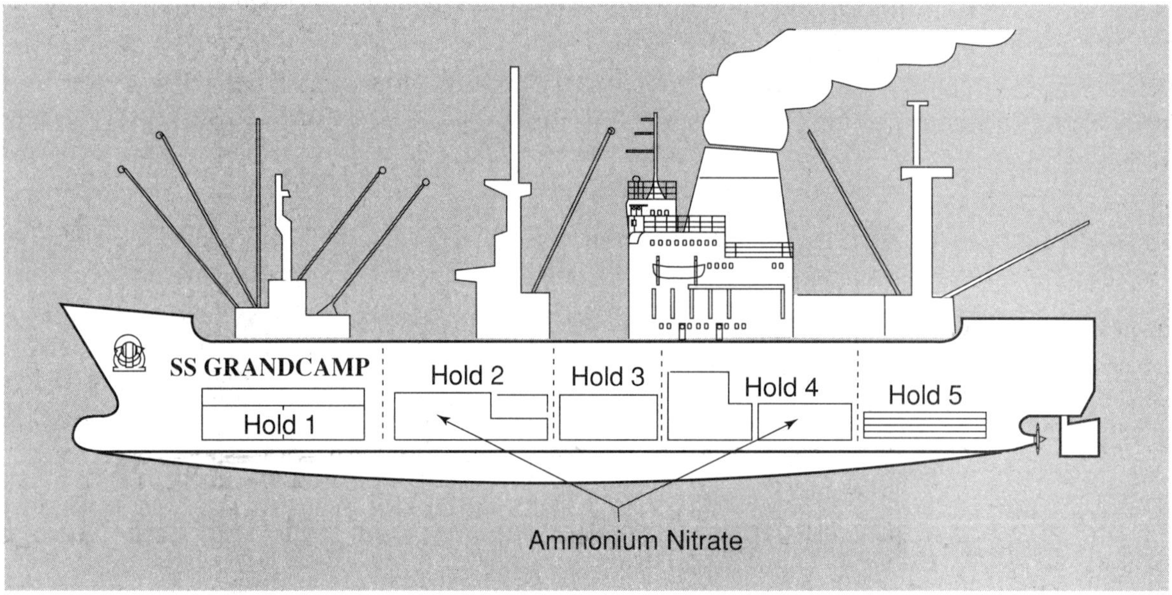

Figure 1.3-1 *SS Grandcamp*

The crew responded to the emergency immediately but was unable to control the fire. The captain was alerted and the crew was ordered to seal the hatch and start the ship's fire smothering steam system, which was designed to keep the cargo from being water-damaged. This decision would end up being the worst possible thing the captain and crew could have done. Instead of smothering the fire, the following conditions occurred within the hold of the ship:

- increased pressure and temperature from the fire and steam started an exothermic reaction in the hold
- the steam initiated thermal decomposition of the ammonium nitrate
- combustible gases were given off by the smoldering fertilizer
- the hatch covers blew off as hot gases belched out of the hold

At 8:45, the Texas City volunteer fire department started spraying water on the fire. The twenty-seven team members fought furiously to control the fire but the ship was so hot that the water vaporized upon contact with the ship. The captain gave the order to abandon ship.

The brilliant orange color of the fire drew a crowd. Explosions and fires were common in the boom town of Texas City. Many of the 18,000 residents had spent hours watching various chemical company fires. Kids from several local high schools cut class and headed to the wharfs to watch the fire. The fire also attracted many of the local chemical plant workers. The Republic Oil Company began shuttling people down to watch the fire including the plant manager, assistant manager, lab manager, and transportation manager. At the nearby Monsanto Chemical Company, employees began to gather to watch the fire from a large drafting room that afforded an excellent view of the burning ship. To most of the spectators, this was a very exciting event. None of the chemical plant employees, bystanders, school children, ship's crew, or airplane pilots circling the area expected the nightmare that unfolded next.

Figure 1.3-2a *Texas City – Volunteer firefighters before the explosion.*

Figure 1.3-2b *Texas City – SS Grandcamp shortly before the explosion.*

Figure 1.3-2c *Texas City – SS Grandcamp at the Port of Houston.*

At 9:12 a.m., the SS Grandcamp unexpectedly and violently exploded. The force of the blast was so great that it killed 581 bystanders, the ship's crew, and the volunteer fire fighters. It knocked two airplanes out of the air, killing their pilots and injuring 3500 other people.The explosion was heard over 150 miles away and sent a mushroom cloud up 2000 feet. It also launched the SS Grandcamp's 1.5 ton anchor two miles away and sent a tidal wave inland that propelled vessels and debris 100 to 200 feet inland. Everything within 1500 feet of the blast was totally destroyed, equalling 700 million dollars worth of property and equipment. Burning fragments of the Grandcamp, which were scattered all over the area, set fire to the Monsanto plant. Additionally, the High Flyer was blown from its moorings. Like the SS Grandcamp, the High Flyer was also loaded with 961 tons of ammonium nitrate. At 1:11 a.m. the High Flyer caught fire, exploded, and sunk a nearby ship, the Wilson B. Keene. Luckily, the area had been evacuated so casualties were light.

Approximately 880 tons of ammonium nitrate had exploded in the hold of the SS Grandcamp. This amount can be contrasted with the 2.5 tons of ammonium nitrate that was used in the

Figure 1.3-3a *Texas City – Harbor after the explosion.*

tragic Oklahoma City blast; the Texas City blast was 300 times more powerful than the Oklahoma City bomb. Fifty years later, the residents of Texas City still remember running north towards the blue sky and away from the billowing black eruptions that rocked Galveston Bay. As the city was rebuilt from the ashes of the explosion, a new era of chemical awareness was ushered in.

Many of the companies that were heavily damaged retained most of their employees and were back up and operational within two years. This spirit of cooperation and communication between the industry and the community has lead directly to the development of CAER programs and allowed Texas City to rebound economically.

Because of the Texas City explosion, wartime studies on the conditions that could destabilize ammonium nitrate were released. Also, emergency response procedures and communication between industry and community were improved. A new awareness of chemical hazards was ushered in, CAER was established, and a catastrophic model existed for future reference.

Figure 1.3-3b *Texas City – The Wilson B. Keene was destroyed when the SS Grandcamp exploded 600 feet away.*

Figure 1.3-3c *Texas City – Propeller pin from the SS Grandcamp.*

1.4 BHOPAL, INDIA, VAPOR RELEASE

During the early morning hours of December 3, 1984, the chemical processing industry was rocked as a large vapor cloud composed of methyl isocyanate (MIC) escaped from the Union Carbide India Limited (UCIL) plant and drifted over the sleeping town of Bhopal, India. (Methyl isocyanate is a major component used in the production of Sevin and Temik, a pesticide used to control insect infestations and crop production in India.) By the time the sun came up, 1400 people had died from the toxic vapor. Seven years later, the official body count had risen to 3800 and those with disabling injuries reached 11,000. Some estimates indicate that over 200,000 people were hurt during this incident. For the first time in modern chemical manufacturing history, an event had taken place near a populated area that demonstrated how vulnerable a community is to the hazards that exist inside the chemical processing industry.

Figure 1.4-1 is a simple flow diagram of the Union Carbide India Limited process that caused the disaster. The process was designed to produce a range of pesticides and herbicides prod-

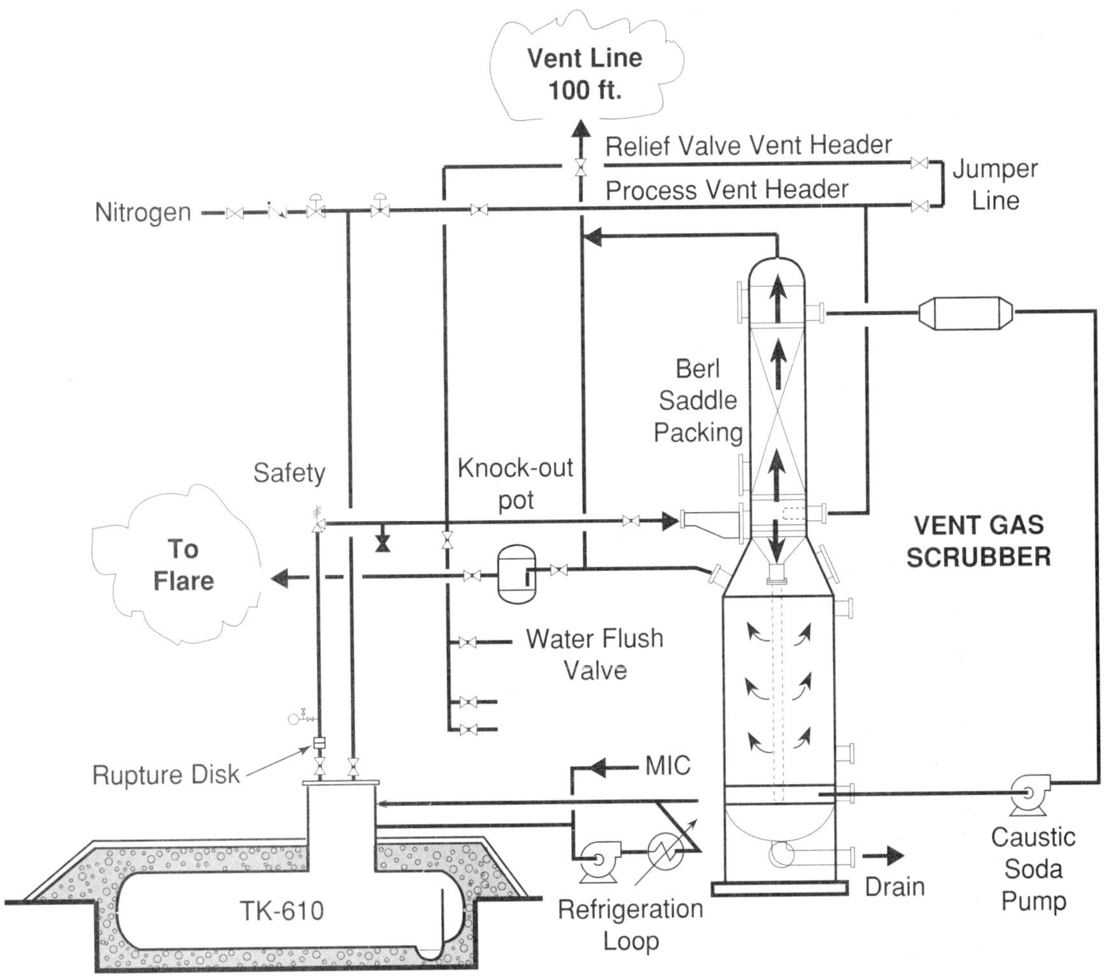

Figure 1.4-1 *Bhopal India, Union Carbide MIC Plant*

ucts essential to the economy of the Indian government. The first step was to produce intermediate phosgene caused by the reaction of carbon monoxide with chlorine. The second step was to mix monomethylamine with phosgene to yield methyl isocyanate (MIC). The third step was to produce carbaryl caused by a reaction of alpha napthol with MIC. (Carbaryl is the end product used in various concentrations to produce a variety of herbicides and pesticides.) Each of the chemicals used in this process is hazardous and requires specific safety handling procedures.

The UCIL process used three 40-foot tanks, each with a storage capacity of 15,000 gallons. The primary tanks were numbered 610 and 611; tank 619 was reserved as an emergency spare and was supposed to remain empty so that in the event of an emergency, the contents of 610 or 611 could be transferred to it.

Plant management decided to shut down the operation on November 14, 1984 in order to perform routine maintenance and to reduce chemical inventories. Plant start-up was scheduled for November 26. Problems within the UCIL process facility became apparent when the process operators started going through their plant's start up-procedures. Vital safety systems and their back-ups were either faulty or had been shut down during the two-week maintenance operation. Four of these critical systems included the vent gas scrubber, flare, MIC tanks instrumentation, and refrigeration unit. (The flare is a device designed to safely burn hazardous emissions before they leave the plant while the vent gas scrubber was designed to neutralize hazardous gases by reacting them with caustic soda). According to plant operators, the temperature, pressure, and level alarms had not been working for several years. Plant management also had dismantled the MIC refrigeration system that was designed to keep the contents of tanks 610, 611, and 619 cool. Instead of repairing this equipment and ensuring that it was operational, management instructed the start-up team to pressure up tank 610 with nitrogen. Tank 610 had approximately forty-two tons of MIC stored in it while tank 611 held about forty tons of MIC. When the operators attempted to pressure up tank 610 with nitrogen they were unable to achieve transfer pressure. A leak was suspected on tank 610's nitrogen outflow or blowdown valve which was probably accurate since tanks 610, 611, and 619 remained under a 2 psig nitrogen blanket during the shutdown. This precaution was necessary since MIC reacts violently with water. Instead of fixing the valve, management decided to switch to tank 611, making a fatal mistake.

At 9:30 p.m., process technicians were water-washing four downstream MIC relief valve vent header (RVVH) gas lines. Tanks 610, 611. and 619 were protected from water contamination by a series of safety valves and rupture disks, an isolation valve, and a slip blind. During this incident the slip blind was not installed. While washing out the lines, the technicians noticed that the water going into the header was not coming out of the open bleeder valves at the same flow rate. It was determined that two bleeder valves were partially clogged and two were completely clogged. The decision was made to keep water-washing. Water began to accumulate in the upper sections of the RVVH, rising as high as twenty feet above the MIC tanks.
(It should be apparent at this point that safety took a back seat to production in the UCIL plant. From an experienced technician's point of view, this is a very unusual way to start-up a hazardous process.)

Another key safety violation was management's decision to install a jumper line between the RVVH and the process vent header. This jumper header is illustrated in figure 1.4-1. The jumper line appears to provide an access point for the water into tank 610.

As the water and iron entered the tank, a number of chemical reactions took place. According to chemists, MIC and water react violently after twenty-three hours of exposure at 60°F. The time frame and incidents do not support that this reaction was the first one to occur. According

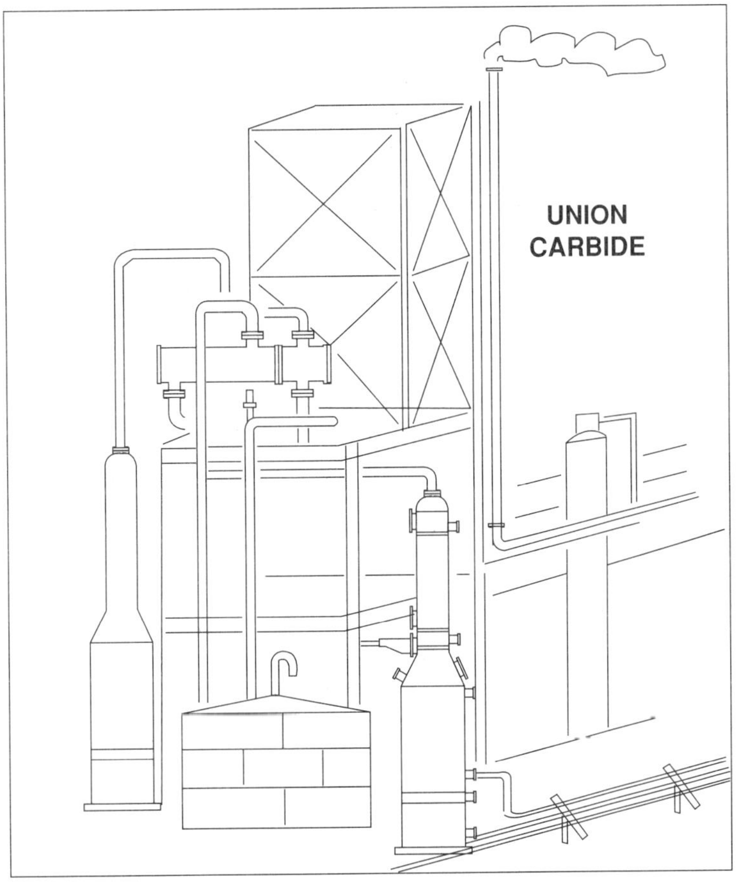

Figure 1.4-2 *MIC plant*

to Union Carbide Corporation's (UCC) piping design specifications, the header should have been made out of stainless steel. Instead it was made out of a cheaper material, namely carbon steel. The reason stainless steel is recommended is because iron (rust or corrosion) acts like a catalyst and will cause MIC to react with itself. When the wash water entered the tank, the exothermic reaction started.

Process technicians reported that the ground became hot around tank 610 as pressures rose and the runaway exothermic reaction reached temperatures of 725°F. Normal gas flow rates into the scrubber averaged 180 kg/hour at 95°F. During the runaway reaction, gas rates exceeded 36,000 kg/hour. Under these runaway conditions, a fully operational flare, scrubber, and refrigeration system would have been overwhelmed by the release.

After the incident, the following occurred:

- UCIL sent technical experts to assist in removing methyl isocyanate and to investigate the incident.
- The Indian Central Bureau of Investigation sealed the plant and limited the corporation's access to the incident site, documents, and employees.
- All UCIL aid packages for victims were rejected by the government.

Several other key factors were uncovered by the investigative teams. The plant had been losing money because of a weak market for Sevin. Plant reductions had decreased operational teams from eleven process technicians and one supervisor to five process technicians and one supervisor. Work loads were redistributed and people were being assigned to work in areas they were unfamiliar with. Maintenance staffs were reduced from six to two technicians. Plant workers were encouraged to take early retirements, 150 employees were placed in a temporary worker pool, and three hundred temporary workers were laid off. As a result, plant moral at the UCIL plant was at an all time low.

However, contrary to the results of an independent investigation, UCC believed that the catastrophe occurred because a disgruntled employee had connected a water hose to tank 610 in an effort to ruin the product. In theory, this employee would have used a crow's foot adapter to make the connection and would have removed the device before the investigative teams arrived. Several operators remembered seeing a water hose in this area.

The UCIL was an operation run entirely by India's government who had a national interest in promoting local control of plant operations and self-sufficiency. In 1984, UCIL was celebrating fifty years of operation. UCIL owned and operated fourteen plants which were split into five divisions employing over 9000 workers. The UCC owned 50.9% of the company's stock; government-run insurance companies owned 24% of the shares; the rest was owned by 23,000 shareholders. The Bhopal plant was a multicultural joint venture in India between UCC and UCIL. Most of the Indian workforce had over ten years of experience with methyl isocyanate operation. UCC provided capital, equipment, and technical expertise to run the plant. The Indian government felt that it was important for the western influence to be removed and to allow their eastern counterparts to run the plant. The last American citizen had left the plant two years earlier.

Summary of Mistakes

- The vent gas scrubber and flare system were undersized and not operational.
- The refrigeration system was dismantled.
- Tank 610's alarms and instrumentation were faulty.
- A jumper line was installed.
- Stainless steel was not used in areas where it was specified for use.
- Plant lay-offs and lack of training caused low morale among employees.
- A safe zone around the plant was not established. Instead, the Indian government and UCIL allowed a densely populated shanty town to grow up around the plant. Most of these people were in the direct path of the release and died first.

- Communication between UCC, UCIL, and the Indian government broke down.
- The Indian government seized control of the plant and arrested the leader of the UCC investigation team. The team was denied access to the plant and employees.
- Official plant documents were altered after the vapor release.
- Five million dollars intended for the victims of the release was withheld by the government and sent back to UCC.
- A two million dollar project funded by UCC to aid the victims of the release was destroyed by the Indian government.
- Samples of the MIC drum were difficult to come by and delayed the analysis phase of the project.

The Indian government filed a suit against the UCC and was awarded 470 million dollars. The government wanted three billion dollars but settled with UCC. A few months later, the government changed hands and the court's first ruling was thrown out.

The initial focus of the Bhopal incident was on the gas release and the people who died. The second phase of the incident resulted in a legal mess between UCC and the Indian government. The third phase of the incident was the "Disaster at Bhopal" detective story which asked What happened? and Who is responsible? The fourth phase of the incident focused on the multicultural differences between the United States and India. The final phase is the evaluation of the disaster. What did we learn? What did we do about what we learned?

Outcomes

The Bhopal tragedy focused on and elevated public concern about storing, transporting, and manufacturing hazardous chemicals near populated areas. The incident alarmed the chemical processing industry and initiated the enactment of laws requiring companies to notify the community and government about hazardous chemicals they use or produce. Bhopal helped refine the Environmental Protection Agency's Federal Superfund Reauthorization and Emergency Planning Councils, a program designed to teach communities how to respond to environmental disasters.

1.5 HOUSTON—PHILLIPS AND ARCO EXPLOSIONS

The chemical processing industry is typically characterized as safe and responsible. Chemical manufacturers have, for the most part, demonstrated a safety conscious atmosphere. In 1980, the CPI as a whole began the process of restructuring. New and innovative technologies were brought into the plants. The rapid assimilation of this new technology was a difficult process for most facilities. In some ways, the traditional safety systems were compromised as bigger and better "toys" were brought into the plants. Older workers had some difficulty adjusting to this new age of computer automation.

Because of this new technology, many companies could produce a lot more with fewer people. Instead of laying off the old workforce, most chemical manufacturers allowed their numbers to be reduced through attrition. When employees retired or resigned they were not replaced. Work loads, schedules, and shift designs were changed. Many plants left the old 8-hour rotating shift in favor of the new 12-hour rotating shift.

Along with this new technology came an era of specialization. The old refinery processes had clearly fragmented into new chemical technologies:

- refinery processes (nineteen main processes with variations of each)
- gas processes
- petrochemical processes (forty-one main processes with variations of each)

Research into synthetic fuel alternatives and petrochemical processes was accelerated during the late 70s and early 80s. A significant number of new patent technologies were developed. And, as the chemical processing industry upgraded its technology, it scrambled to put in place adequate safety systems. The term adequate does not necessarily mean good. Yet, when employees were asked how they thought their employers prioritized safety, quality, and production, most technicians ranked their company's priorities as:

1. PRODUCTION !!!
2. quality
3. safety

Chemical manufacturers were attempting to keep safety near the top of their company's priority list, however the necessary synergy was not there to put it at the top. Because of this, a number of work practices existed in the industry which led to catastrophe. Some of these work practices included:

- the industry's use of unskilled, unqualified contract labor to perform jobs in the plant
- industry chemical hazards were not regulated
- prevention of catastrophic releases was not regulated

Phillips Chemical Company in Houston produced high density polyethylene for use in a number of plastic-related specialties. During the early 80s, a lot of seasoned employees had retired from Phillips. The knowledge base and experience levels dropped significantly and a mechanism did not exist for capturing the technical expertise of this group and disseminating it into the existing work force. Very few of the people who retired were replaced; work assignments were shifted and redistributed to the remaining employees. Fewer people were doing a lot more.

In August of 1989, two workers were accidentally killed inside the plant. A heavy cloud hung over the Phillips complex as they attempted to uncover what they needed to do to prevent this from happening again. Unfortunately; the system that was in place at Phillips and at many other companies was responsible for the high-risk environment.

On October 22, 1989, employees at Phillips were asked to remove a polyethylene blockage to a loop reactor. The employees refused the assignment because hydrocarbon vapors were heavy in the area and they believed a valve was leaking through. Phillips employees were willing to perform the job if the company would shut down the operation and secure the area. The company made a decision to bring in contract workers and have them remove the polyethylene restriction while the unit was kept on line.

The difficulty of the operation became apparent as the contract crew attempted to complete their assignment. After working on the assignment for many hours, the decision was made to bypass one of the unit's safety features in order to speed up the work. A critical valve in the procedure had two high pressure hoses connected to it in order to help position the valve. Unknown to the contract employees, the high pressure hoses were installed backwards and the valve could not be closed without switching the hoses. When the valve was opened, 85,000 pounds of high pressure gases dumped from the reactor system. A mixture of ethylene, hexane, hydrogen, and isobutane dumped to the concrete mat. Within a few minutes, a deadly vapor cloud engulfed the plant.

Contract workers and maintenance staffs fled the site as the vapor cloud grew. The Phillips operational staff responded immediately to the emergency. While others fled, they remained in their area frantically attempting to stop the release. When the explosion occurred, twenty-three members of the operational staff were killed and the entire plant was destroyed. The explosion registered 3.5 on the Richter scale as debris was scattered over a six mile radius and damage estimates ranged from 750 to 800 million. OSHA sited Phillips for 566 willful safety violations.

Eight months later, on July 5, 1990, the ARCO chemical plant in Sheldon, Texas was having trouble with a safety system on a waste water tank. The problem had been going on for several months and seemed to have finally reached its peak. When the tank exploded it leveled the immediate area and killed seventeen men. It was later determined that two gauges could have prevented the accident, but one was broken and the other was blocked out.

Most process technicians view OSHA as a reactive rather than proactive organization. Government officials usually show up *after* someone has been killed or a complaint has been filed. Over a ten month period, forty-two men had been killed within a six mile radius. National attention had been focused on the Houston area. OSHA and the EPA were determined to investigate the incidents, identify the problems, and put a plan in place which would prevent future catastrophic releases of toxic, reactive, or flammable chemicals.

Outcomes

The Phillips and ARCO incidents directly led to the development of the Process Safety Management (PSM) standard. The PSM standard poured new life into existing programs such as Hazard Communication and HAZWOPER. OSHA and the EPA called into question the industry's use of contract employees. One of the largest tremors experienced by the PSM standard was the creation of industrial partnerships between education, government, and industry. This alliance formed the foundation upon which the community college and university process technology program was built.

SUMMARY

Process technicians work in chemical plants and refineries, the pulp and paper industry, food processing plants, power plants, city utilities, and many other areas. According to the American Chemical Society, in 1994 there were over 500,000 process technicians in the United States. Process operators fall into an educational category referred to as process technology. Process technology is defined as the study and application of the scientific principles associated with the operation and maintenance of the chemical processing industry. The roles and responsibilities of process technicians include knowing the equipment and technology they are working with, being able to maintain and operate this equipment, complying with safety and environmental regulations, applying quality techniques to operational processes, collecting and analyzing samples, and troubleshooting system problems. The process operator of the future will be required to have a formal education credential and advanced training prior to taking an entry level position in the chemical processing industry. To obtain this type of advanced training a person will need to complete a two-year degree program in process technology which is offered at many community colleges and universities. The Associate of Applied Science (AAS) or one-year certificate verifies that basic training has been completed.

REVIEW QUESTIONS

1. In your own words, describe the explosion of the SS Grandcamp.

2. Describe the hazards of handling, storing, and transporting ammonium nitrate.

3. Describe the special conditions required to safely handle methyl isocyanate. What could have been done to prevent the UCIL accident?

4. In your opinion, who was responsible for the Bhopal tragedy? Please assign blame to UCC, UCIL, ICBI, and/or plant technicians.

5. If a large quantity of methyl isocyanate was released over your home town tonight, what would happen?

6. Why were so many people killed during the SS Grandcamp explosion?

7. Who was primarily responsible for the Phillips explosion? Consider the management, contractors, and the operational staff.

8. Who was primarily responsible for the ARCO explosion? Consider the management, contractors, and the operational staff.

9. List the mistakes that occurred during the Bhopal, Texas City, Phillips, and ARCO incidents. What is common between them?

10. A storm is moving into your area while you are performing a minor line up. The last step in your assignment is to open V-1 and close V-2 on top of a 60 foot distillation column. Just as you start up the ladder it begins to rain. Describe what you would do next.

11. You have been told that tank 107 is unsafe and leaking to the top soil and should be used only under emergency conditions. Four hours into your shift, laboratory results indicate that the biological system is not functioning. Tank 107 is the only available hold-up tank. What do you do?

12. After completing the six week company training program, you are assigned to shift team B for on-the-job training. As you are filling out your checklist you notice that several pressure gauges are not working right, the area is old and noisy, numerous leaks are visible, and the equipment is in need of a paint job. Describe your first entry in your unit logbook.

Chapter 2

Basic Principles of Safety

OBJECTIVES

- *List the basic principles of safety.*
- *List the general safety rules used by the plants.*
- *Explain the principles of accident prevention.*
- *List the principles of accident investigation.*
- *Identify the methods of hazard recognition.*
- *Explain the Occupational Safety and Health Act.*
- *Describe government mandated training for process technicians.*

KEY TERMS

Accident—an unplanned disruption of normal activity resulting in an injury or equipment damage.

Code of Federal Regulations (CFR)—contains all of the permanent rules and regulations of OSHA and is produced in paperback format once a year.

Designated Equipment Owner— the process technician who operates a piece of equipment or process.

Hazard communication (HAZCOM)— standard known as the workers right to know.

Hazardous waste operations and emergency response (HAZWOPER)—a government standard enacted by the Occupational Safety and Health Administration to ensure the safety of industrial workers. HAZWOPER provides a structured procedure for hazardous waste cleanup and emergency response.

National Institute for Occupational Safety and Health (NIOSH)—one of three primary agencies formed under the Occupational Safety and Health Act. It is primarily responsible for toxic substance handling criteria, safety and health issues, and making policy for new health standards.

Occupational Safety and Health Administration (OSHA)—one of three primary agencies formed under the Occupational Safety and Health Act. OSHA is the primary agency that investigates industrial accidents, inspects workplaces, and establishes workplace safety and health standards and penalties.

Occupational Safety and Health Review Commission (OSHRC)—one of three primary agencies formed under the Occupational Safety and Health Act. This review commission conducts hearings on contested issues, assesses penalties, investigates incidents, supports, modifies or overturns OSHA findings.

Permit System—a regulated system that uses a variety of permits for various applications. The more common applications are cold work, hot work, confined space entry, opening blinding, permit to enter, and lockout/tagout.

Personal protective equipment (PPE)

Potable water—drinkable water.

Respiratory Protection—a standard designed to protect employees from airborne contaminates.

The Federal Register—a publication that provides information on current OSHA standards & shows all adopted amendments, deletions, insertions & corrections to government standards.

Unsafe act—any act that increases a person's chance of having an accident.

Unsafe condition—a condition in the working environment that increases a technician's chance of having an accident.

2.2 GENERAL PLANT SAFETY RULES

General safety rules are designed to protect human life, the environment, and physical equipment, or facilities. Before entering a refinery or chemical plant a simple overview of the general plant safety rules is conducted. This includes:

- Do not go to a fire, explosion scene, accident, or area exposed to a vapor release unless you have specific duties or responsibilities at that scene.
- Obey all traffic rules.
- Do not park in designated fire lanes.
- Report injuries immediately.
- Stay clear of suspended loads.
- Smoke only in designated areas of the plant.
- Drink from designated water fountains and potable water outlets only.
- Use the right tool for the right job.
- Report to the designated equipment owner before entering an operating area.
- Stay in your assigned area.
- Do not use illegal drugs or alcohol in the plant.
- Do not use firearms or cameras in the plant.
- Take steps to remove hazardous conditions.
- Review and follow all safety rules and procedures including personal protective equipment, hazard communication, respiratory protection, permit system, hazardous waste operations and emergency response, housekeeping, and fire prevention.
- Know and understand the alarms and rules associated with vapor release, fire or explosion, evacuation, and "all clear."

2.3 ACCIDENT PREVENTION

The basic principles of accident prevention include a safe working environment, safe work practices, and effective management. If management does not support the safety program, process technicians do not follow the rules of a safety program, or the environment is riddled with hazards, the system will not work.

Plant managers are the central focus of safety program management. Managers design the company's safety policy, enhance safety awareness, and facilitate safe and healthful work conditions. In the typical chain of command, safety coordinators are the designated specialists who provide current safety supervision and expertise on the program. The safety coordinator takes much of the day to day operation of a large safety program off the plant manager. The safety specialist receives training in specialized areas and carefully analyzes and compares existing and new governmental rules and guidelines to current plant operations.

The first line supervisor has the greatest impact on a plant's safety program. These individuals typically have the most knowledge about a specific process and are looked upon as the primary contact for safety awareness. Typically, team safety meetings are conducted by the first line supervisor. First line supervision provides the bridge between operations personnel and the safety supervisor or, in some cases, the plant manager.

Process technicians ensure that the rules of the safety program are maintained by all of the people entering the facility. Technicians are the backbone of any safety program because they are responsible for:

- good housekeeping
- hazard recognition
- safe work practices
- safe work environment

EFFECTIVE MANAGEMENT

- Accept Responsibility for Safety
- Write Safety Policy
- Stimulate Safety Awareness
- Use Engineering, Administrative and PPE controls
- Teach Dept. Heads and Give Authority
- Utilize First Line Supervisors
- Hold Stewardship Interviews
- Integrate Safety Department into Corporate Culture

SAFE WORK PRACTICES

- Conduct Job Safety Profile
- Develop Job Standards
- Develop Standard Operational Procedures
- Integrate HAZCOM, HAZWOPER, PSM, Respiratory Protection, PPE, DOT, Permit System, Fall Protection, Hearing Conservation, Environmental Standards and other selected topics

SAFE WORKING ENVIRONMENT

- Develop Corporate Safety Philosophy
- Implement General Plant Safety Program
- Launch Accident Prevention Program
- Launch Hazard Recognition Program
- Implement Plant Training

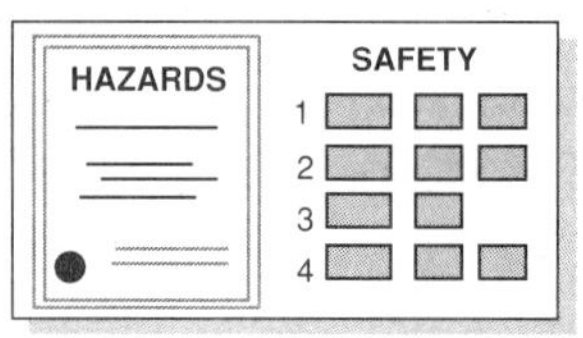

Figure 2.3-1 *Accident Prevention*

2.4 ACCIDENT INVESTIGATION

Accident investigation is designed to identify the point of failure and prevent its reoccurrence. The point of failure typically falls into two groups: hazardous conditions or poor work practices.

First line supervisors:

- conduct accident and near accident investigations
- visit the accident site as soon after the accident as possible
- take photographs
- collect data and information from witnesses and those involved in the accident
- fill out accident reports, including analyzing what has happened and identifying objectives

2.5 HAZARD RECOGNITION

In a typical safety program, "hazard recognition" is considered to be a management function. In this text, hazard recognition is considered to be a process technician's primary function. Accidents are typically defined as an unplanned disruption of normal activity resulting in a bodily injury or equipment damage. Controlling the number of accidents that occur in a plant is a factor associated with workers' performances, equipment and tools, and the working environment. The key to accident prevention is linked to determining the cause and preventing its reoccurrence. By understanding the causes of accidents, process technicians can more easily defend against their occurrences. Accidents can be broken down into four parts:

1. the accident (struck by object)
2. immediate causes (did not use PPE)
3. contributing causes (poor instructions)
4. the consequences of the accident (fatality)

Figure 2.5-1 *Hazard Recognition*

Classification words for accidents include: struck by, struck against, caught in, on, or between, fall from above, fall at ground level, strain or overexertion, electrical contact, and burn.

Unsafe acts and unsafe conditions are responsible for many of the accidents that occur inside the plant. An unsafe act is defined as any act that increases a person's chance of having an accident. Unsafe acts can be eliminated by a no tolerance policy. Since this is not the most effective way to do business, other alternatives need to be researched. Unfortunately, the negative reinforcement theory makes it very difficult to convince someone to stop performing unsafe acts. Many of these people think the short cuts they take are worth it or that "It can't happen to me." Unsafe conditions are easier to handle than unsafe acts. An unsafe condition is defined as a condition in the working environment that increases a technician's chance of having an accident. Unsafe conditions can be corrected simply by encouraging process technicians to identify and correct an unsafe condition. Incentive programs and training have proven to be an effective tool in preventing unsafe conditions.

2.6 Occupational Safety and Health Act (OSHA)

In 1970, a landmark piece of legislation was passed that required the chemical processing industry to make safety and health on the job a matter of federal law. The Occupational Safety and Health Act (OSHA), brought about sweeping changes that affected 4 million American businesses, and more importantly, 57 million employees and their families. In 1969, 2.5 million disabling injuries and 14,000 deaths were directly linked to safety and health violations. The purpose of OSHA is to:

1. remove known hazards from the workplace that could lead to serious injury or death
2. ensure safe and healthy working conditions for American workers

The coverage of the legislation is extensive in scope. OSHA applies to four broad categories: agriculture, construction, general industry, and maritime.

There are three primary agencies responsible for the administration of OSHA:

1. **National Institute for Occupational Safety and Health (NIOSH)**
   This agency is responsible for safety and health research. It also
   - develops toxic substance handling criteria
   - researches safety and health issues
   - recommends new safety and health standards

2. **Occupational Safety and Health Administration (OSHA)**
   - investigates catastrophes and fatalities
   - establishes new safety and health standards and penalties
   - inspects workplaces

3. **Occupational Safety and Health Review Commission (OSHRC)**
   This independent agency does the following?
   - conducts hearings for situations on noncompliance that are contested
   - can access penalties, conduct investigation, support, modify or over turn OSHA findings

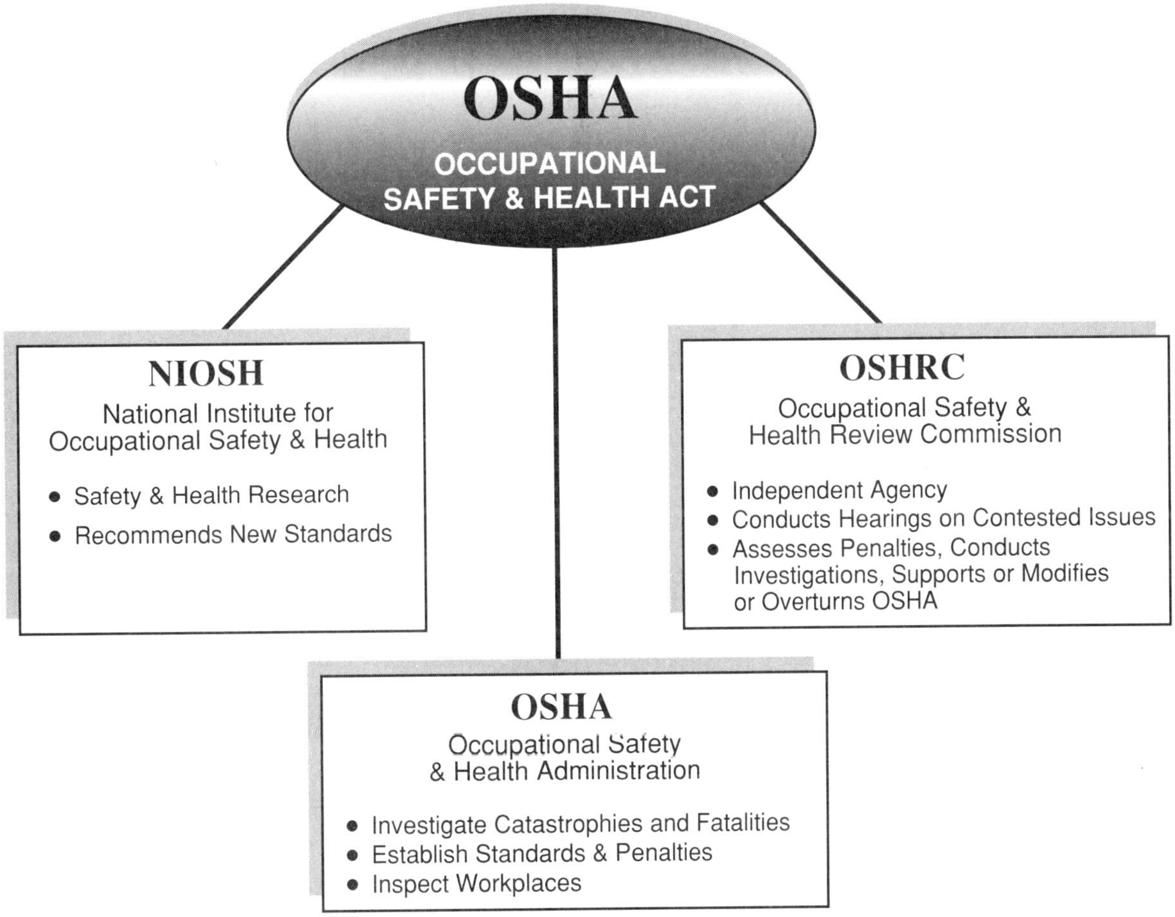

Figure 2.6-1 *OSHA*

The best source for information on OSHA standards is the *Code of Federal Regulations* (CFR). The CFR contains all of the permanent rules and regulations of OSHA and is produced in paperback format once a year. Another publication that provides information on current OSHA standards is *The Federal Register. The Federal Register* is a publication that shows all adopted amendments, deletions, insertions, and corrections to government standards. This document is published daily and provides information to the public on federal activities.

When an OSHA inspection is conducted, prior warning is not given to the plant. The following steps are typically followed:

1. Opening conference—an inspector meets with the employer and explains the purpose of the visit. The inspector is accompanied by a company representative on the plant walk-through tour. The inspector is required to meet with a reasonable number of employees.
2. Inspection tour—the inspector reviews plant injury and illness records and reviews operations. It is a good idea to fix any problem that the inspector notices on the spot. This shows the inspector that the company is serious about conforming to safety standards.
3. Closing conference—the inspector notes any observed violations.

2.7 MANDATED TRAINING

Government mandated training for process technicians covers a variety of topics and includes formal classroom and hands-on training that takes place upon initial assignment to a process unit annually, every two years, or every third year. The following list includes most of the essential, mandated training topics and reference numbers. This is not the entire list; however, it does include most of the common topics taught between industrial facilities.

- **Hazard Communication** (HAZCOM) 29 CFR 1910.1200—required upon initial assignment.
- **Hazardous Waste Operations and Emergency Response** (HAZWOPER) 29 CFR 1910.120—awareness, operations, required upon initial assignment and annually each year thereafter.
- **Bunker Gear** 29 CFR 1910.120—required upon initial assignment and annually each year thereafter.
- **Respiratory Protection** 29 CFR 1910.134—required upon initial assignment and annually each year thereafter.
- **Personal Protective Equipment** 29 CFR 1910.133 & 135—required upon initial assignment.
- **The Control of Hazardous Energy** (lockout/tagout) CFR 1910.147—required upon initial assignment and every two years thereafter.
- **Process Safety Management** (hot work)- 29 CFR 1910.119—required upon initial assignment.
- **Permit Required Confined Spaces** 29 CFR 1910.146—required upon initial assignment.
- **Fire Extinguisher** 29 CFR 1910.157—required upon initial assignment and annually each year thereafter.
- **Department of Transportation** (DOT) 29 CFR 171-177.
- **Occupational Noise Exposure** 29 CFR 1910.95—required upon initial assignment and audiometric testing each year thereafter.

- **Scaffold User Safety Inspections** 29 CFR 1926.450-.454—required upon initial assignment.
- **Benzene** 29 CFR 1910.1028—required upon initial assignment.
- **Electrical Training for Unqualified Persons** 29 CFR 1910.332—required upon initial assignment and every two years thereafter.
- **Asbestos Awareness and Gasket Removal** 29 CFR 1910.1200/1001 and 1926.59/1101—required upon initial assignment and annually each year thereafter.
- **Specifications for Accident Prevention Signs and Tags** 29 CFR 1910.145—required upon initial assignment.
- **Bloodborne Pathogens** 29 CFR 1910.1030—required upon initial assignment.
- **Powered Industrial Trucks** 29 CFR 1910.178—required upon initial assignment.

SUMMARY

Before a new process technician enters a refinery or chemical plant, a simple overview of the general plant safety rules is conducted. Process technicians ensure that the rules of the safety program are maintained by all of the people entering the facility. Technicians are the backbone of any safety program because they are responsible for:

- good housekeeping
- hazard recognition
- safe work practices
- safe work environment

The basic principles of accident prevention include a safe working environment, safe work practices, and effective management. Accidents are defined as an unplanned disruption of normal activity resulting in an injury or equipment damage. Determining the cause of an accident and preventing its reoccurrence are the keys to accident prevention. By understanding the causes of accidents, process technicians can more easily protect against their occurrence. An unsafe act is defined as any act that increases a person's chance of having an accident.

The three primary agencies responsible for the administration of the Occupational Safety and Health Act are the National Institute for Occupational Safety and Health, the Occupational Safety and Health Administration, and the Occupational Safety and Health Review Commission.

REVIEW QUESTIONS

1. Describe the purpose and the key elements of HAZCOM.
2. Describe the purpose and the key elements of HAZWOPER.
3. Give an example of PPE worn by a process technician.
4. What does the term "designated equipment owner" mean?
5. Why is respiratory protection important to a process technician?
6. What is potable water?
7. Define an industrial accident.
8. Contrast an unsafe act with a safe act.
9. There are three primary agencies responsible for the administration of the Occupational Safety and Health Act. List them.
10. What is NIOSH?
11. Describe the role of OSHA.
12. What is the best source for information on OSHA standards?
13. What is accident investigation designed to do?
14. The point of failure of an accident typically falls into two groups. List them.
15. Describe the critical components of hazard recognition.
16. Contrast the roles of the plant manager, first line supervisor, and process technician in creating safe and healthy work conditions.
17. List five government mandates that pertain to a process technician.

Chapter 3

Process Safety Management

Government Mandate:

Process Safety Management 29 CFR 1910.119.
Training required upon initial assignment.

OBJECTIVES

- ***Describe the process safety management (PSM) standard.***
- ***Explain the written procedures requirement of the PSM standard.***
- ***Identify the critical components of process safety information.***
- ***List the seven employee training issues contained in the PSM standard.***
- ***List the four issues covered under the employee participation section.***

KEY TERMS

Compliance audits—conducted by OSHA auditors to ensure compliance with governmental rules and regulations.

Emergency response plan—a written plan that documents how specific individuals should respond during an emergency situation.

Environmental Protection Agency (EPA)—established by Congress in the 60s to oversee the protection of our environment. The agency monitors air and water quality, waste disposal, chemical safety and more than twelve laws enacted by Congress.

Flow diagram—a simplified process drawing that uses standard symbols and diagrams to identify equipment and flows.

Hot work—welding, cutting, or using a spark-producing device.

Mechanical integrity—applies to the soundness of a plant process.

Process hazard analysis—designed to identify the causes and consequences of fires, vapor releases, and explosions.

Process safety management (PSM) standard—designed to prevent the catastrophic release of toxic, hazardous, or flammable materials that could lead to a fire, explosion, or asphyxiation.

3.2 THE PSM STANDARD 29 CFR 1910.119

After the Bhopal incident in 1984 and the Phillips plant and ARCO explosions in 1989 and 1990, OSHA and the EPA went to work on a new standard that would limit the possibility of these events happening again. After a tedious investigation into the causes of various industrial explosions, fires, and vapor releases, the government issued the Process Safety Management standard. This document was modeled after the API 750 industry consensus standard. Key elements of the PSM standard include:

- employee participation
- process safety information
- process hazard analysis
- operation procedures
- employee training
- emergency planning and response
- pre-startup safety review
- mechanical integrity
- hot work permit
- management of change
- incident investigation
- contractors
- compliance audits
- trade secrets

The PSM standard has had a widespread, systematic impact on the chemical processing industry. From the moment the PSM "Final Rule" was implemented in 1992, a change occurred that restructured the occupation of the process technician. (The term *final rule* is used to indicate that the standard has been implemented and is now in force.) Job descriptions have been rewritten and the process technician of the future looks much different from the one of the 1970s and 80s.

Since 1992, industrial manufacturers have initiated partnerships with their local community colleges and universities. The traditional apprentice training function has gravitated from industrial training managers towards the regionally accredited halls of academia. This transition has saved the chemical processing industry millions in training budget dollars. Company training managers and coordinators are now able to focus on site specific, on-the-job training programs and government mandated safety and environmental training. The site-specific training function will never move into the community college or university environment since there is such a variety of industrial processes. Only a handful of colleges have pilot plant facilities and bench-top units.

This transition can be documented all across the United States as certain elements of the PSM standard require a higher level of certification. This higher educational credential and need for

EMPLOYEE PARTICIPATION
- Written Program
- How Employees will Access Hazard Identification System
  - identify hazards
  - gather information
  - communication system

OPERATIONS PROCEDURES
- Operations and Maintenance
- Reflect current work practices
- Process Properties
- Hazards
- Start-up, Shutdown
- Change of Chemicals

PROCESS SAFETY
- Process Flow Diagram
- Equipment, Process Description, Limitations
- Consequences of Deviation
- Safety and Relief Devices
- Electrical Classifications
- Characteristics of Chemicals
- Process Chemistry
- Mixing Chemicals

EMPLOYEE TRAINING
- Process Overview
- Training Records and Method
- Attendance & Competency
- Training materials reflect current work practices
- Control access to unit
- Refresher training
- Contractors must inform and train their employees and document that training

Figure 3.2-1 *The PSM Standard*

some training redundancy has changed the occupation of the process technician from little if any formal training to the preferred status of a two-year process technology A.A.S. degree. Standardized college curricula exist in Texas and Louisiana with several other states positioning themselves for educational standardization.

The four sections of the PSM standard that have had the greatest impact on process technician apprentice training are employee participation, process safety information, operation procedures, and employee training. It is virtually impossible for a new apprentice technician to successfully address these sections of the PSM standard without formal training.

3.3 EMPLOYEE PARTICIPATION

Employees need to be involved in planning how they as well as contractors will participate in the process hazard analysis. This team effort will include employees from all levels. Access to information needs to be readily available to all employees and contractors.
The employee participation section has four important parts:

1. identifying hazards
2. gathering crucial process and hazard information
3. communicating crucial process and hazard information
4. the written procedure and how employees access it

3.4 PROCESS SAFETY INFORMATION

The process safety information section includes information on the accidental mixing of chemicals and the products they produce, potential hazards, and safe handling procedures. This section provides an accurate assessment of the physical data associated with chemicals: toxic, reactive, corrosive, and explosive limits of specific substances. A simplified flow diagram of the process should be included in this section of the standard so that a new person can easily see the various steps of the process. The complete list of the items that should be included in the process safety information include:

- process flow diagram
- equipment and process description
- operational limits of equipment
- consequences of deviations
- safety and relief system design
- electrical classifications
- physical characteristics of chemicals: toxic, reactive, corrosive, and explosive limits
- process chemistry
- inadvertent mixing of chemicals under a variety of conditions, including temperature, pressure, and flow rates

3.5 PROCESS HAZARD ANALYSIS

The process hazard analysis section is designed to identify the causes and consequences of fires, vapor releases, and explosions. All of the hazards on a unit need to be identified. During the start-up, check for new equipment or a major process change. The equipment and construction must conform to established standards. Potential hazards and risks should be identified and discussed prior to implementation. All of the operational procedures, safety procedures, and training must be in place prior to start-up.

3.6 OPERATION PROCEDURES

Written procedures fall into two specific categories: operating and maintenance. Written operating procedures apply to process technicians and must accurately reflect current work practices, process properties, hazards, process change of chemicals, start-up, and shut-down procedures.

Maintenance procedures apply to maintenance employees and must identify the department's collective consensus for how a specific task should be done. The chemicals that are used in the process should be well documented in the maintenance procedure. The mechanical integrity of the equipment must be ensured and any deficiencies identified and repaired. Any change in the process or equipment requires written procedures and training before any work can be done. An evergreen feature for keeping all operational and maintenance documentation up to date should be built into the procedure system. An example of an "evergreen feature" could be a validation page at the end of each procedure section allowing all who use the procedure to verify its accuracy by making comments and submitting them through a structured system.

3.7 EMPLOYEE TRAINING

OSHA and the EPA believe that the key to preventing catastrophic emergencies inside of the chemical processing industry is adequate employee training. This was the conclusion of the governmental groups that investigated the Phillips Chemical Company and ARCO vapor release and explosions. The employee training aspect of the process safety management standard includes seven sections:

1. process overview
2. training records and method used to administer training
   (document attendance and competency achieved)
3. identify chemicals used in the process
4. control access to and from the process unit
5. training materials must reflect current work practices
6. provide refresher training
7. contractors must inform and train their employees and document that training

3.8 CONTRACTORS

Under the law, contractors fall under the same requirements of the PSM standard as company employees. This text applies all of the requirements of the standard to contract employees.

3.9 PRE-START-UP SAFETY REVIEW

The pre-start-up safety review and process hazard analysis work together during the evaluation of the start-up process. During this phase, the equipment and construction are carefully scrutinized. A risk assessment is made as each area is studied. It is during this phase that operational and safety procedures are written and tested.

3.10 MECHANICAL INTEGRITY

Mechanical integrity is a term that applies to the soundness of a plant process. Mechanical integrity is monitored during the hazard analysis and pre-start-up section of this standard and during operational rounds and routine maintenance.

3.11 HOT WORK PERMIT

The primary purpose of the hot work permit is to protect technicians from fires and explosions that could result from hot work that is performed in their area. Hot work is defined as welding, cutting, or using a spark-producing device. Specific written procedures must be in place to comply with this section of the standard.

3.12 MANAGEMENT OF CHANGE

The PSM standard requires that a formal system be in place that makes deviation from standard operational procedures very structured with layers of approval required. This system must be documented and the proper training set in place for each employee.

3.13 INCIDENT INVESTIGATION

Incident investigation is a procedure that takes place when a near catastrophic or catastrophic event occurs. This part of the PSM standard has three parts:

1. assemble incident investigation team within forty-eight hours
2. address all of the findings
3. correct all of the action items identified during the investigation

3.14 EMERGENCY PLANNING AND RESPONSE

Emergency planning and response is a complex plan that impacts the company and community. Within the company are several levels of response. The first category deals with emergency plans that impact plant employees, what they do, designated meeting points, and key contacts. The next level deals with people who are responding to the emergency including their roles and responsibilities. The next level deals with the community, the plant, and the emergency. The PSM standard requires that these sections be documented and appropriate training be completed. Written action plans are required.

3.15 COMPLIANCE AUDITS

OSHA inspectors will verify compliance of each section of the PSM standard. A typical compliance audit has an opening meeting and a walk-through to observe site conditions. The auditor will require documentation for a records review and will select one or more areas to conduct employee interviews. A closing meeting will be held to discuss the findings of the compliance audit.

3.16 TRADE SECRETS

Many companies have specific trade secrets that need to be in place to protect them from their competitors. This section of the standard allows companies to protect themselves by entering into confidentiality agreements with their employees.

SUMMARY

The PSM standard has had a significant impact on the process technician and the chemical processing industry. From the moment the PSM "Final Rule" was implemented in 1992, a change occurred that restructured the occupation of the process technician. The key elements of the PSM standard that have had the greatest impact on process technician training include process safety information (including the process flow diagram, equipment and process description, operational limits of equipment, consequences of deviations, safety and relief system design, electrical classifications, physical characteristics of chemicals, process chemistry, and inadvertent mixing of chemicals under a variety of conditions), process hazard analysis (designed to identify all of the hazards on a unit, the causes and consequences of accidents, operational procedures, safety procedures, and training), operation procedures, and employee training. The government believes that the key to preventing catastrophic emergencies inside of the chemical processing industry is adequate employee training. Upon initial assignment, employees will go through a process overview. This will require an understanding of symbols and diagrams, equipment, and typical processes. Training materials must reflect current work practices and operational procedures. Refresher training will occur based on government mandates. Contractors must inform and train their employees and document that training.

REVIEW QUESTIONS

1. Describe the purpose of the PSM standard.
2. What is emergency response?
3. Define hot work.
4. Describe the impact of the PSM standard on process technicians and the chemical processing industry.
5. What does the government believe the key to preventing catastrophic emergencies inside the chemical processing industry?
6. What, in your opinion, are the three most important aspects of the process safety management standard?
7. Compare Section 3.7, Employee Training to traditional apprentice training programs at your local college.
8. Review Section 3.4, Process Safety Information and identify each area that is covered in a college process technology program.

Chapter 4

Hazard Communication

Government Mandate:

Hazard Communication 29 CFR 1910.1200

OBJECTIVES

- *Describe the hazard communication standard.*
- *Identify the physical properties and hazards associated with handling, storing, and transporting chemicals.*
- *Describe physical and health hazards associated with exposure to chemicals.*
- *Describe the key elements of a material safety data sheet.*
- *Describe a hazardous chemical inventory list.*
- *Explain the purpose of a written hazard communication program.*
- *Identify methods used to protect process technicians from hazardous chemicals.*
- *Identify safety signs, tags, and warning labels utilized by process technicians.*
- *Describe toxicology and the terms associated with it.*
- *Describe the material classification system for DOT.*
- *Describe the Hazardous Materials Identification System.*
- *Describe the National Fire Prevention Association labeling system.*

KEY TERMS

Acute effect—when a chemical has an immediate adverse effect on biological tissue.

Ambient—surrounding or encircling.

Bulk containers—have a rated design for liquids of 119 plus gallons, 882 pounds for solids, and a water capacity greater than 1000 pounds for gases.

Ceiling level "C"—the maximum allowable human exposure limit for an airborne substance.

Central nervous system—the brain and spinal cord.

Chemical Transportation Emergency Center (Chemtrec)—provides information on chemicals. It is a service of the Chemical Manufacturers Association and can be reached by calling 1-800-424-9300, 24-hours a day.

Chronic effect—a slow developing adverse effect (such as cancer) that may take ten to twenty years to appear. Typically results after long-term exposure to a chemical.

Cutaneous hazards—a chemical that affects the dermal layer of the body.

Dermatitis—skin irritation or inflammation.

Flammable gas-—any gaseous material that forms an ignitable mixture when combined with 13% air at 68°F and 14.7 psi.

Flammable liquid—a flashpoint below 100°F (37°C).

Flashpoint—the lowest temperature at which a flammable liquid will produce a rich enough vapor concentration to ignite in the presence of an ignition source.

Hazardous chemical—a chemical that has been determined to be a physical hazard to humans.

Health hazard—a chemical that has been statistically proven by one or more scientific studies to have acute or chronic health risks for humans.

Hematopoietic system— the blood forming system in the human body.

Hepatotoxin—a chemical suspected of causing liver damage.

Incompatible—a term applied to chemicals that react violently when they come into physical contact with each other.

Lethal concentration 50 (LC_{50})— when the concentration of an airborne chemical reaches the level where it kills 50% of the test animals.

Liquefied compressed gas—a liquid/gas mixture that has a gauge pressure of 40 psig at 70°F (21.1°C).

Lower explosive limit (LEL)—the lowest concentration at which a vapor or gas will produce a rich enough vapor concentration in air to ignite in the presence of an ignition source.

Milligrams per liter (mg/l)—milligrams of material per liter of air.

Nephrotoxin—a chemical suspected of causing kidney damage.

Neurotoxin—a chemical suspected of causing nerve damage. Some linkages exist between neurotoxins and behavioral and emotional abnormalities.

Non-bulk containers—have rated capacities for liquids (119 gallons or less), solids (882 pounds or less or total capacity of 119 gallons), and gases (a water capacity of 1000 pounds or less).

Non-flammable compressed gas—a material that does not conform to the definitions of a flammable or poisonous gas and exerts an absolute pressure of 41 psi at 68°F.

Olfactory—a term used or associated with the sense of smell.

Parts per million (ppm)—Parts of material per million parts of air.

Permissible exposure limit (PEL)—established by OSHA. Determines how long and at what concentration a worker can be exposed to a chemical.

Physical hazard—a chemical that statistically falls into one of the following categories: combustible liquid, compressed gas, explosive or flammable, organic peroxide, oxidizer, pyrophoric, unstable, or water reactive.

Primary hazard—the hazard classification of the material with the greatest risk percentage component being shipped. (See 172.101 table for CFR.)

Subsidiary hazard—material other than primary hazard.

Target organ toxin—a chemical that selectively targets a specific organ in the body.

Upper explosive limit (UEL)—the highest concentration at which a vapor or gas will produce a rich enough vapor concentration in air to ignite in the presence of an ignition source.

4.2 THE HAZARD COMMUNICATION PROGRAM 29 CFR 1910.1200

The purpose of the Hazard Communication Standard is to ensure that the hazards associated with the handling, transport, and storage of chemicals in a plant are evaluated and transmitted to affected personnel. A fundamental principle of the chemical hazard communication program is that informed technicians are less likely to be injured by chemicals and chemical processes than uninformed technicians. According to the standard, all of the chemical inventories and processes within a chemical plant or refinery must be evaluated for potential hazards and risks. Where a risk is found, essential information and training is required for the existing workforce and new employees.

Requirements of the Standard

Since the chemical processing industry manufactures chemicals and employs technicians, it is responsible for both sections of the OSHA standard that address chemical manufacturers and user responsibilities and employer requirements. Chemical manufacturers are required by the HAZCOM standard to:

- analyze and assess the hazards associated with the chemical and develop written procedures for evaluating chemicals
- document the hazard and develop material safety data sheets and warning labels
- disseminate the information to affected individuals
- label, tag, and attach warning documentation to chemicals leaving the workplace

Employers are responsible for the development of a written hazard communication program, a hazardous chemical inventory list, and associated material safety data sheets. This written program should be designed so that it is given to the new employee upon initial assignment. The materials should be site specific, readily accessible by plant personnel, and include an evergreen feature that will keep it up to date. Employers are also required to provide training to employees about the hazards of the chemicals they will be working with, how to read a material safety data sheet (MSDS), how to select and use personal protective equipment, and how to read and use one of the three standard labeling systems: Department of Transportation (DOT), Hazardous Materials Identification System (HMIS), and National Fire Protection Association (NFPA).

Delivery of the Standard to Employees

The HAZCOM standard requires employers to provide information or training to employees about their plant's hazard communication program. Fundamental information that must be provided to a process technician includes the key elements of the HAZCOM standard, their plant's written hazard communication program, a detailed hazardous chemical inventory list, and associated material safety data sheets along with warning labels, tags, and signs. Information should be included on how to access the HAZCOM system, the chemical inventory list, and MSDS's 24-hours a day, seven days a week. Employers are required by law to provide open access to HAZCOM materials. This is why the HAZCOM standard is frequently referred to as the workers' right to know act.

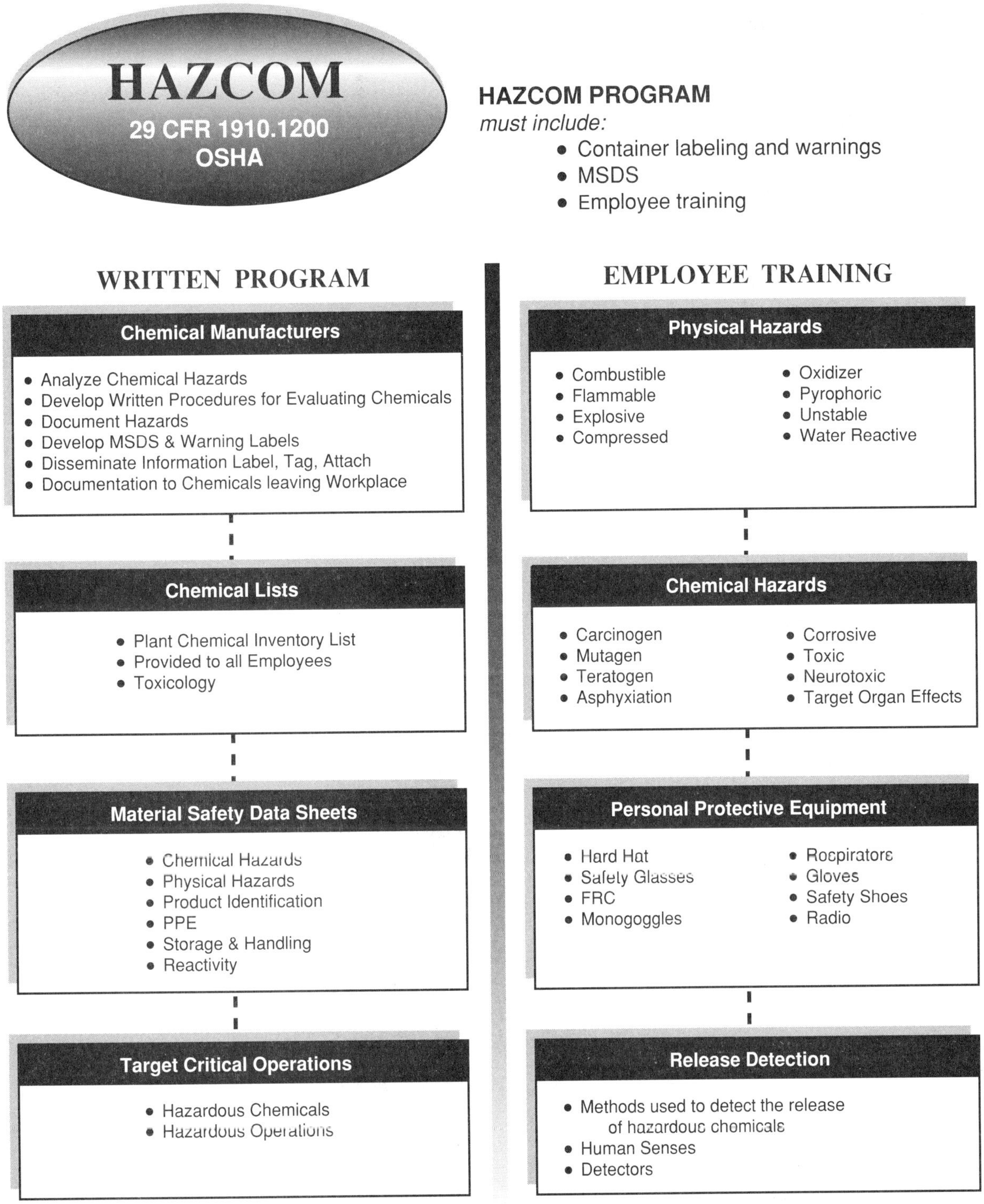

Figure 4.2-1 *HAZCOM*

The chemical processing industry initiates the delivery of HAZCOM training upon initial assignment to the plant. Training focuses on the physical and health hazards associated with exposure to chemicals. Additional information is provided on the toxicology, physical properties, and hazards associated with handling, storing, and transporting chemicals. New technicians are required to review company procedures used to protect employees from hazardous chemicals and specific operations are identified that may expose an employee to a chemical. The training section also includes the selection and use of personal protective equipment and the methods and observations utilized to detect the release of hazardous chemicals.

4.3 SAFELY HANDLE, STORE, AND TRANSPORT HAZARDOUS CHEMICALS

Transporting, storing, and handling chemicals requires that a process technician understand the systems, equipment, and technology they are working with, the physical hazards associated with chemicals in their facility, the health hazards associated with chemicals in their facility, and chemical routes of entry into the human body. The process technician must use the material safety data sheets and proper labeling, signs, and tags.

In order to safely store and transport a chemical, a technician should be familiar with the system they are working with. These systems vary from plant to plant so it would be difficult to identify all of the industrial processes that are available in the petrochemical, gas processing, and refinery industry. Examples of these industrial processes include alkylation, isomerization, catalytic cracking, hydrodesulfurization, propylene, polyethylene, benzene, crude oil distillation and over a hundred more. New technicians are carefully trained on these systems over a specified period of time. New technicians are carefully monitored and are not released to work alone until they have passed the rigorous requirements established by each plant. Site specific training is a critical part of transporting, handling, and storing chemicals.

During the apprentice training program, most technicians become familiar with the basic equipment and technology found in the petrochemical, gas processing, and refinery industry. Although the processes that this equipment is used in may change, the equipment remains constant. Most apprentice training programs cover valves, piping, vessels, pumps, compressors, steam turbines, heat exchangers, cooling towers, fired heaters, boilers, distillation columns, reactors, extruders, flare systems, environmental awareness, safety training, math, chemistry, physics, roles and responsibilities of process technicians, working in self-directed work teams, quality control, basic process principles, process instrumentation, symbols and diagrams, industrial processes, basic hand tools, and minor maintenance training.

4.4 PHYSICAL HAZARDS ASSOCIATED WITH CHEMICALS

A physical hazard is a chemical that falls into one of the following categories:

- combustible liquid—has a flashpoint between 100°F (38.8°C) and 200°F (93°C)
- compressed gas—has a gauge pressure of 40 psig at 70°F (21.1°C)
- explosive—a chemical characterized by the sudden release of pressure, gas, and heat when it is exposed to pressure, high temperature, or sudden shock
- flammable gas—forms a flammable mixture with air at ambient temperature

- flammable liquid—has a flashpoint below 100°F (37°C)
- organic peroxide—explodes when temperature exceeds a specified point
- oxidizer—a chemical that promotes combustion in other materials through the rapid release of oxygen, usually resulting in a fire
- pyrophoric—a chemical that ignites spontaneously with air at temperatures below 130°F (54.4°C)
- unstable—a chemical that will react (condense, decompose, polymerize, or become self-reactive) when it is exposed to temperature, pressure, or shock
- water reactive—a chemical that reacts with water to form a flammable or hazardous gas

Gas/ Vapor	Lower-Upper Flammable Limit (vol%)		Autoignition Temperature (C)	Boiling Point (C)
Acetone	2.6	13.0	465	
Acetylene	2.5	100	305	-83.6
Ammonia	15.0	28.0	-	-33.5
Benzene	1.3	7.9	560	80.2
1,3-Butadiene	2.0	12.0	420	
n-Butane	1.8	8.4	405	
Carbon Monoxide	12.5	74.0	-	-191.5
Ethane	3.0	12.4	515	-88.3
Ethyl Alcohol	3.3	19.0	365	78.3
Ethylene	2.7	36.0	490	-103.8
Toluene	1.2	7.1	480	110.6
Gasoline	1.3	7.1	440	
n Heptane	1.05	6.7	215	
n-Hexane	1.2	7.4	225	
Hydrogen	4.0	76.0	400	
Methane	5.0	15.0	540	-161.4
Methyl Alcohol	6.7	36.0	385	64.7
Propylene	2.4	11.0	460	
n-Pentane	1.4	7.8	260	
Propane	2.1	9.5	450	-44.5

Figure 4.4-1 *Physical Properties of Chemicals*

4.5 HEALTH HAZARDS ASSOCIATED WITH CHEMICALS

- Carcinogens—known cancer causing substances. Carcinogenicity labels indicate that a substance may cause cancer upon contact or have a chronic effect over a large number of years. For example, coal tar is known to cause cancer on an exposed body part. Vinylchloride is known to have a chronic or long-term effect on the liver when it is inhaled over a long period of time. Most items listed on the MSDS will indicate whether a substance is a suspected carcinogen. Most carcinogen labels indicate that the substance has been proven to have an impact on laboratory animals.

- Mutagen—a chemical that is suspected to have the properties required to change or alter the genetic structure of a living cell.
- Teratogen—a substance that is suspected to have an adverse effect on the development of a human fetus.
- Reproductive toxin—a chemical that inhibits the ability of a person to have children. Chemicals are routinely tested for this property.
- Asphyxiation—simple and chemical. Simple asphyxiation occurs when oxygen is removed or displaced by a chemical. For example, a sudden release of nitrogen gas into an enclosed space will dissipate life supporting oxygen causing loss of consciousness and death. Chemical asphyxiation occurs when a chemical blocks or impedes the ability of a person to use oxygen. For example, carbon monoxide attaches to red blood cells and prevents oxygen from being utilized even if it is available in sufficient quantities.
- Anesthetic—dulls the senses. Example: alcohol.
- Neurotoxic—slows brain down. Example: lead and mercury.
- Allergic response—a person may not have an allergic reaction immediately to a chemical. In some cases, it takes multiple exposures. Example: poison ivy.
- Irritants—a chemical that causes temporary discomfort when it comes into contact with human tissue.
- Sensitizers—a chemical that effects the nerves. Example: Phenol will be absorbed through the skin and will sensitize the affected area.
- Corrosives—a chemical that causes severe damage to human tissue. Example: sulfuric acid.
- Toxic—a chemical that has been determined to have an adverse health impact. This term is linked to toxicology, dose, and response and at least one experiment that has indicated the chemical has an adverse effect on laboratory animals.
- Highly toxic—a chemical that requires only a small amount of exposure to be lethal.
- Target organ effects—a chemical that contacts the body at a remote location and is transferred to another area of the body where it has an adverse effects on a specific organ.

Hazardous chemicals can enter the body in four ways: inhalation, absorption, ingestion, and injection.

1. **Inhalation**—airborne chemicals enter the body through the mouth or nose and may irritate the nose, throat, bronchi, and deep lung tissue. Some gases or vapors will not irritate the respiratory tract but will be absorbed into the blood system through the lungs.This process can effect the blood, brain, bone, liver, fatty tissue, kidney, and colon.

Hazardous chemicals can alter healthy cells and they may become cancerous. Airborne chemical particles may be retained in the respiratory tract and cause allergic reactions, lung scaring, cancer, or fibrosis.

2. **Absorption (skin contact)**—only a small number of chemicals are known to have serous impacts upon skin contact. These chemicals should be identified quickly if they are in your unit. When a hazardous chemical comes into contact with skin, four reactions are possible:
    1. nothing will occur
    2. chemical will react with skin and cause primary irritation
    3. chemical will penetrate the skin and cause sensitization
    4. chemical will penetrate the skin and enter the blood

3. **Ingestion**—a process where a hazardous chemical is transmitted to food, liquid, or cigarettes and then swallowed. Airborne chemicals also may be taken in through the respiratory system and then swallowed in the saliva or mucous. Typically, large quantities of hazardous chemicals are not ingested; however, small quantities of arsenic, lead, or mercury can be highly toxic over a period of time.

4. **Injection**—a process where a toxic or hazardous material is injected into the body by a needle or sharp object or through cuts in the skin.

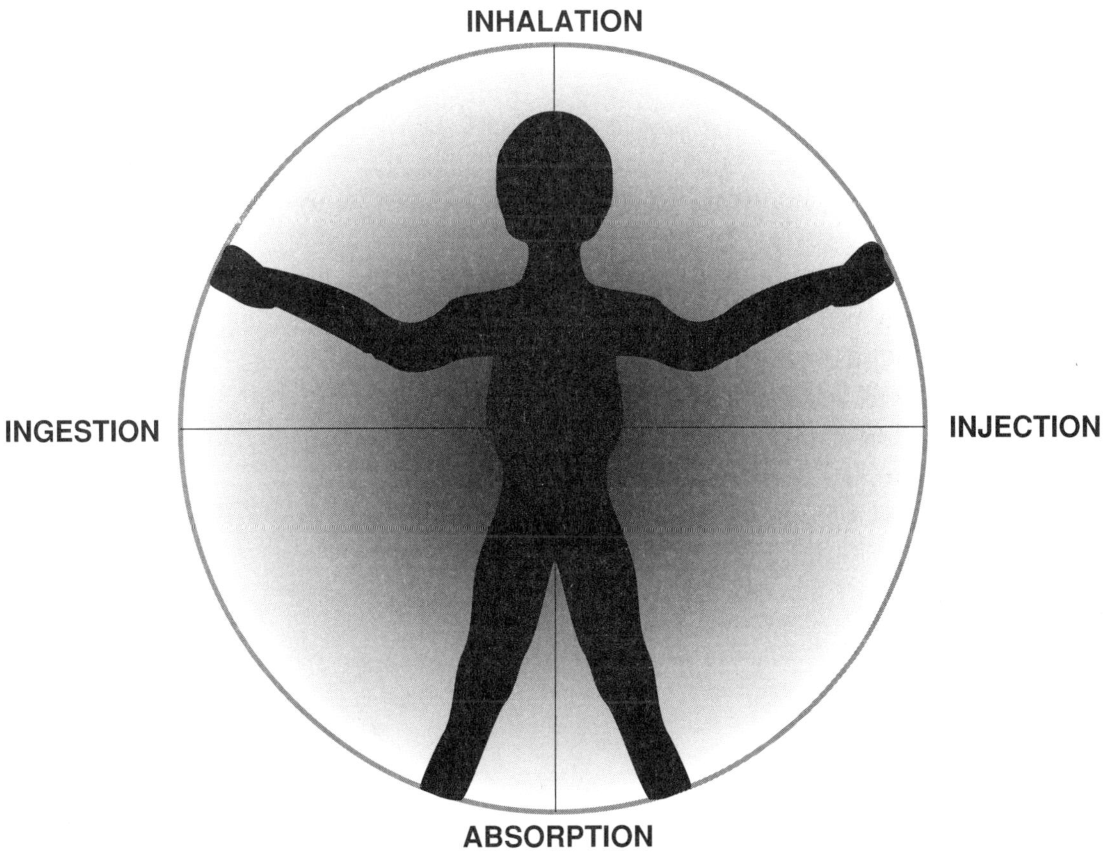

Figure 4.5-1 *Routes of Entry*

4.6 THE MATERIAL SAFETY DATA SHEET (MSDS)

It has been estimated that one out of every four workers in the United States handles a chemical. While most chemicals are not harmful in small doses, caution should be exercised when the dose (amount) exceeds the recommended exposure (time) limits. The development of the material safety data sheet (MSDS) is the responsibility of the chemical's manufacturer. A typical MSDS has nine or ten sections:

1. **SECTION ONE:** Product Identification and Emergency Information—chemical name, chemical family, formula, trade name, company code, manufacturer address, and an emergency telephone number operating 24 hours a day.
2. **SECTION TWO:** Hazardous Ingredients—chemical component percentages and OSHA hazard.
3. **SECTION THREE:** Health Information and Protection or Hazards Identification—nature of hazard such as eye, skin contact, inhalation, and ingestion, and first aid. Exposure limits, precautions, and personal protective equipment also covered.
4. **SECTION FOUR:** Fire and Explosion Hazard—flashpoint, flammable limits, autoignition temperature, general hazards, fire fighting procedures, and hazardous combustion products.
5. **SECTION FIVE:** Physical Data and Chemical Properties—vapor pressure, boiling point, evaporation rate, appearance and odor, specific gravity, solubility in water, percent volatile, density, freezing, and melting point.
6. **SECTION SIX:** Spill Control Procedure—land spill procedure, water spill procedure, and vapor release. Methods to control and to protect yourself.
7. **SECTION SEVEN:** Regulatory Information—Department of Transportation shipping description, flashpoint, TSCA information, Comprehensive Environmental Response, Compensation and Liability Act, CERCLA information, and Superfund Amendments and Reauthorization Act (SARA) information.
8. **SECTION EIGHT:** Reactivity Data—stability of material, conditions to avoid instability, materials and conditions to avoid incompatibility, hazardous decomposition of product, and hazardous polymerization.
9. **SECTION NINE:** Storage and Handling—electrostatic accumulation hazard, storage temperature, storage, transport pressure, loading, unloading temperature, and viscosity at loading.
10. **SECTION TEN:** Personal Protective Equipment—respiratory protection, clothing and gloves, face shield, goggles, ventilation, and so on.

MATERIAL SAFETY DATA SHEET

IDENTIFY ***(As used on label and list)***

Section 1
Manufacturer's name, address

Emergency Telephone:
Information telephone:
Date prepared:
Signature of Preparer:

Section 2
Hazardous Ingredients

Hazardous components:
Chemical name:
Exposure limits:
- OSHA PEL, permissible exposure limit
- ACGIH TLV, threshold limit value
- OEL, occupational exposure limit
% of composition:

Section 3
Physical / Chemical Characteristics

Boiling point:
Vapor pressure:
Vapor density:
Solubility in water:
Specific gravity:
Melting point:
Evaporation rate:
Appearance and odor:

Section 4
Fire and Explosion Hazard Data

Flashpoint:
Flammable limits:
Extinguishing media:
Special fire fighting procedures:
Unusual fire and explosion hazards:

Section 5
Reactivity Data

Incompatibility:
Hazardous decomposition or byproducts:
Hazardous polymerization:

Section 6
Health and Hazard Data

Routes of entry:
Health hazards:
Carcinogenicity:
Signs and symptoms of exposure:
Medical conditions generally aggravated by exposure:
Emergency and first-aid procedures:

Section 7
Precautions for Safe Handling and Use

Steps to be taken if material is released or spilled:
Waste disposal methods:
Precautions to be taken in handling and storing:
Other precautions:

Section 8
Control Measures

Respiratory protection:
Ventillation:
Protective gloves:
Eye protection:
Other protective clothing or equipment:
Work / hygienic practices:

Figure 4.6-1 *Sample MSDS*

4.7 TOXICOLOGY

Toxicology is the science that studies the noxious or harmful effects of chemicals on living substances. The fundamentals of toxicology include a relationship between dose and response. Dose is defined as the amount of chemical entering or being administered to a subject while response is defined as the toxic effect the dose has upon the subject.

During a scientific study, a graph is developed which plots dose and response. Statistical analysis is used to determine the lethal dose 50 (LD_{50}). The LD_{50} is the expected dosage that will kill 50% of the lab animals in a controlled experiment. The term LC_{50} is applied when the dosage is applied through inhalation. Both the LC_{50} and LD_{50} are indicators of the toxicity of a specific substance. For example:

LC_{50} .350 grams—least toxic
LC_{50} .150 grams
LC_{50} .005 grams—most toxic

The degree of toxicity of a substance also has a relationship with dose and response. As toxicity increases the required dosage decreases. Some substances are eight million times more toxic than others. Chemicals have four possible routes of entry into the human body: skin absorption, inhalation, ingestion, and injection. For example, when phenol is absorbed through the skin it damages the nerves. The degree of damage depends on the dosage. A small dose may result in a tingling sensation or numbness, a medium dose in partial paralysis, and a large dose in death. Likewise,when benzene is inhaled into the lungs it is carried by the blood cells throughout the system. Small quantities of benzene are harmless while a high LC_{50} would result in bone marrow damage and an inability to produce red blood cells. Also, when potassium cyanide is ingested it destroys living cells.

Exposure limits to various materials allow a process technician to work safely with a chemical over an extended period of time. Exposure limits are numerical values from four different exposure limit categories:-

1. Threshold limit value (TLV)—established standard.
    - time weighted average (TLV-TWA)—based on an eight hour day, forty hours a week without harmful side effects.
    - short-term exposure limit (TLV-STEL)—maximum fifteen minute exposure, four times a day in sixty minute intervals.
    - ceiling (TLV-C)—maximum limit. Do not exceed!
2. Occupational exposure limits (OELs)—developed by OSHA and industrial hygiene to protect workers from hazards when a TLV does not exist.
3. Permissible exposure limits (PEL)—OSHA mandated regulatory limits.
4. Supplier recommended limits.

There are three ways to detect the release or presence of hazardous chemicals in the workplace: bodily senses, monitoring equipment, and industrial hygiene. The first way is by smell, sight, sound, taste, or touch. Technicians should always be alert for unusual vapor releases, strange smells, abnormal sounds and vibrations, and chemical irritants that impact the eyes, ears, and nose. The second way to detect a release is with portable and continuously operated detectors. This process provides a safer layer of protection than relying on the bodily senses alone. The third way to detect chemicals is through industrial hygiene. Each plant should have an industrial hygienist who is trained in the use of detection equipment appropriate for the chemicals used in the facility.

4.8 SAFETY SIGNS, TAGS, AND WARNING LABELS

Petrochemical, gas processing, and refineries all store, transport, and move large quantities of chemicals. Each of these chemicals must be properly labeled and identified. Appropriate safety signs must be displayed so everyone can see them and be located where they are most needed. Tags should be used where needed and a well defined permit system established. Training for each of these areas must be given to all employees. Examples of warning labels, safety signs, and tags include:

- Warning Labels—Chemical information (identify the chemical, hazards, and manufacturer)
- Safety Signs—chemical hazard such as benzene, H_2S, chlorine, eye wash, safety stations, speed limits, and so on
- Tags—cold work permits, opening or blinding permits, hot work permits, permit to enter, lockout/tagout, energy isolation, and unplugging

The chemical processing industry uses three widely recognized hazard communication labeling systems: DOT, HMIS, and NFPA.

4.9 DEPARTMENT OF TRANSPORTATION (DOT) 29 CFR 171-177

The DOT labeling system utilizes a diamond shaped pattern with colors, symbols, and numbers that correspond to specific hazards. These signs are typically used as placards for tank cars and trucks. The DOT numbering system does not indicate the severity of the hazard but rather the hazard's division or classification.

DOT Material Classification

Shipments of hazardous materials are regulated by the DOT. These regulations contain specific information on how hazardous materials are identified, placarded, documented, labeled, marked, and packaged. Hazardous material shipments that are not in compliance with federal regulations will be delayed and can result in severe penalties. In civil cases, marking the wrong name on a container can result in fines up to $25,000 per violation. In criminal cases, fines up to $500,000 and five years in jail can be imposed for *intentionally* shipping a hazardous chemical and attaching the wrong MSDS.

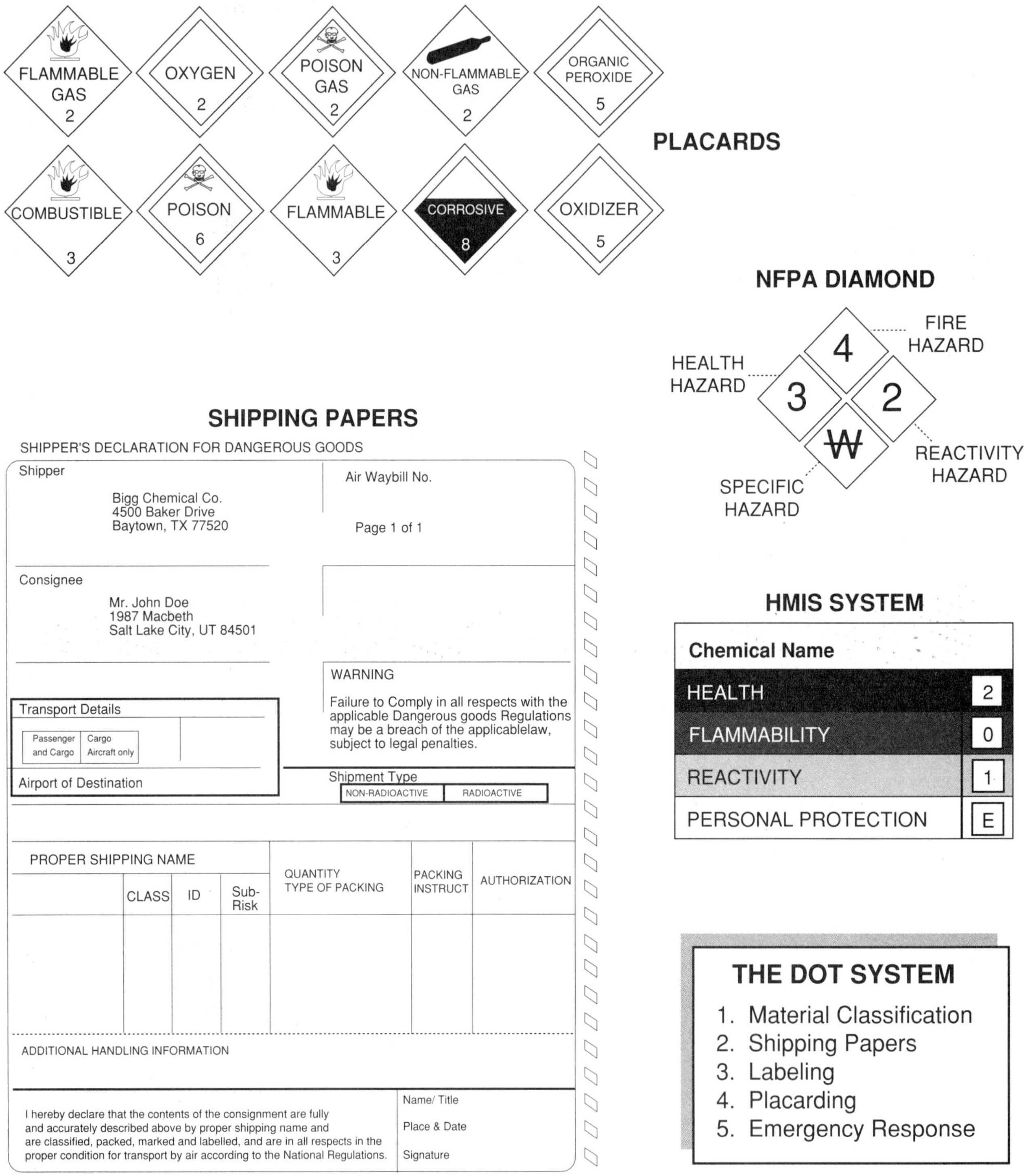

Figure 4.9-1 *DOT System*

Materials are classified for transportation in ten different categories. These classifications include:

1. explosive
2. gases
3. flammable liquids
4. self-reactive
5. oxides
6. poisonous and infectious materials
7. radioactive materials
8. corrosive materials
9. miscellaneous hazardous materials
10. peroxides

DOT Shipping Papers

The shipping papers for a chemical must contain information that will allow easy access and understanding to the material being transported. The information should be in English and may be hand written. Codes and abbreviations can not be used in the official document. The following information should be included in the shipping papers:

- basic description of material
- name and address of shipper
- CHEMTREC number or company 24-hour hotline
- the proper shipping name of material
- hazard class or division (immediate hazards to health)
- UN or NA material classification number
- packing group of material
- basic description such as N.O.S., flammable liquid, and so on (risk of fire or explosion)
- quantity and volume
- precautions to be taken in the event of an accident
- methods for handling fires, spills, and leaks
- first aid procedures

Note: Additional information is needed for radioactive material.
An MSDS should be included with any material being shipped.

DOT Marking, Labeling, and Placarding

The marking and labeling procedure for chemicals that are being shipped must follow specific guidelines. Labels, signs, or tags should be in an unobscured, durable format and in English. In most cases, this information is contrasted against an eye-catching background. Package orientation markings are designed to ensure that the product is stored in the safest possible way.

Hazardous materials must be labeled with primary and subsidiary hazard labels. The primary hazard will take precedence over the subsidiary with the hazard class number being removed from the secondary hazard label. Aircraft cargo must be labeled with an Authorized Cargo Aircraft Only Label. Labels should be placed near the proper shipping name and not on the bottom of containers where they cannot be seen easily.

SHIPPING PAPERS

SHIPPER'S DECLARATION FOR DANGEROUS GOODS

Shipper

Bigg Chemical Co.
4500 Baker Drive
Baytown, TX 77520

Air Waybill No.

Page 1 of 1

Consignee

Mr. John Doe
1987 Macbeth
Salt Lake City, UT 84501

WARNING

Failure to Comply in all respects with the applicable Dangerous goods Regulations may be a breach of the applicablelaw, subject to legal penalties.

Transport Details

Passenger and Cargo	Cargo Aircraft only

Airport of Destination

Shipment Type

NON-RADIOACTIVE	RADIOACTIVE

PROPER SHIPPING NAME	CLASS	ID	Sub-Risk	QUANTITY TYPE OF PACKING	PACKING INSTRUCT	AUTHORIZATION

ADDITIONAL HANDLING INFORMATION

I hereby declare that the contents of the consignment are fully and accurately described above by proper shipping name and are classified, packed, marked and labelled, and are in all respects in the proper condition for transport by air according to the National Regulations.

Name/ Title

Place & Date

Signature

Figure 4.9-2 *Typical Shipping Form*

Placards are used to display the proper shipping name and classification of chemicals. Placards typically contain the following information:

- proper shipping name
- material classification number (at bottom of placard)
- material identification number (in center of placard)

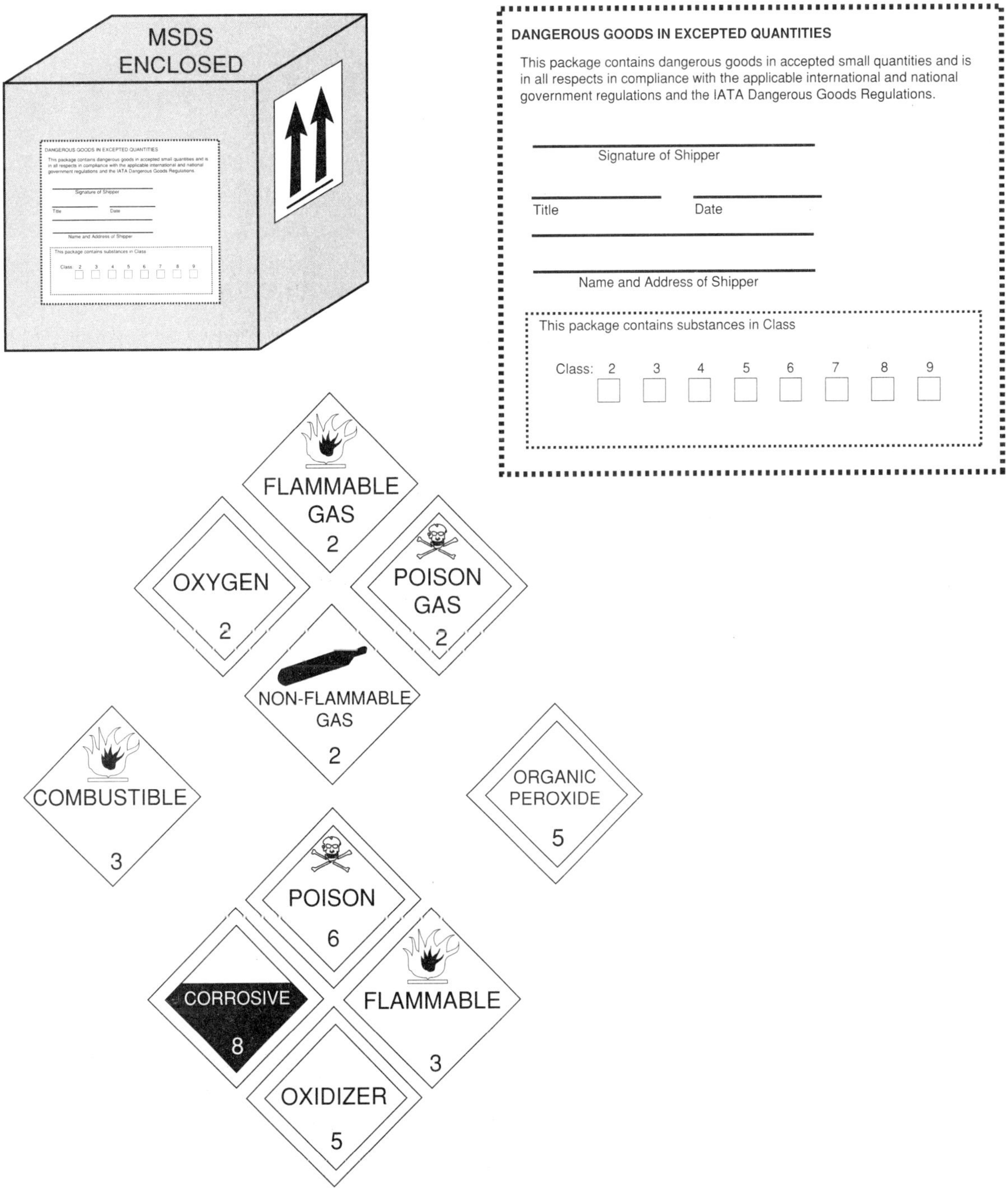

Figure 4.9-3 *Marking, Labeling, Placarding*

Process technicians place placards on all chemicals leaving the physical boundaries of the plant except for infectious substances and combustible liquids in non-bulk packages. Cargo tanks carrying hazardous materials should be placarded on all visible sides. Placards are typically the responsibility of the shipper and are displayed in the diamond pattern.

Hazardous Materials Identification Systems (HMIS)

HMIS is another labeling system frequently used by industrial manufacturers to identify the hazards associated with chemicals. The HMIS communicates its information to process technicians by listing five essential topics: the chemical name, health hazard, flammability hazard, reactivity hazard, and personal protective equipment requirements. Each topic block is color-coded and receives a severity ranking from one to four. The only exception to this rule falls under the PPE topic block where a letter is displayed that corresponds to an index.

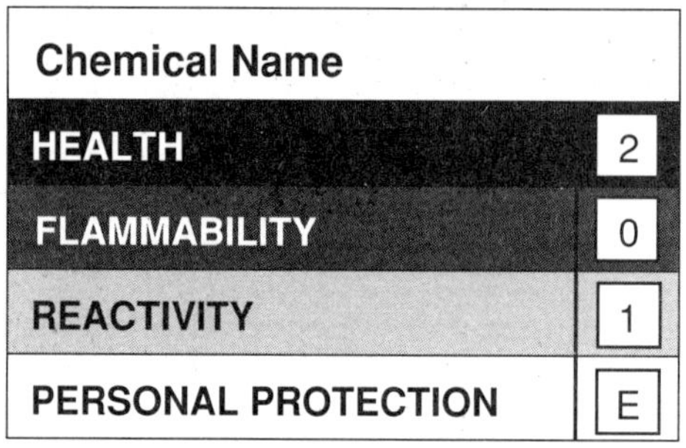

Figure 4.9-4 *HMIS System*

National Fire Protection Association (NFPA)

A third hazard communication labeling system is the NFPA. This system utilizes the four small diamonds arranged so that they form one large diamond. Each small diamond is colorcoded and has its own severity rating from one to four. The four color-coded topic blocks are blue for health hazards, red for fire hazards, yellow for reactivity hazards, and white for specific hazards.

Any unmarked vessel that contains a chemical must be identified. Portable samples should also have the appropriate labels. The law requires any vessel that stores a chemical to be properly marked. Some flexibility exists in this area, so permanent signs may not be required. Some plants use electronic systems to identify the contents of a vessel by its number. The only exception to the labeling rule is if a small quantity of chemical is being moved to another location within the same facility where it will be used immediately. When a chemical leaves a plant it falls under the HAZCOM guidelines and must be labeled properly and, in most cases, be accompanied by an MSDS.

Pipeline transfers fall under the HAZCOM guidelines and the appropriate paperwork is electronically shipped during the transfer process. Chemical transport in trucks falls under HAZCOM and the DOT. Truck drivers are given an MSDS and must display the appropriate placards, labels, and signs.

HEALTH HAZARD

4 - Deadly
3 - Extremely Hazardous
2 - Hazardous
1 - Slightly Hazardous
0 - Normal Material

FIRE HAZARD (Flash points)

4 - Below 73°F
3 - Below 100°F
2 - Below 200°F
1 - Above 200°F
0 - Will not burn

REACTIVITY HAZARD

4 - May detonate
3 - May detonate with heat or shock
2 - Violent chemical change
1 - Not stable if heated
0 - Stable

SPECIFIC HAZARD

Oxidizer	OXY
Acid	ACID
Alkali	ALK
Corrosive	COR
Use no water	₩
Radiation Hazard	☢

NFPA DIAMOND

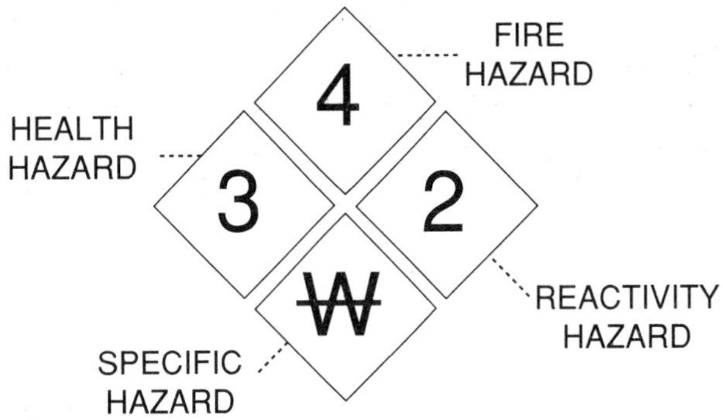

- ☐ FIRE HAZARD
- ☐ HEALTH HAZARD
- ☐ REACTIVITY HAZARD
- ☐ SPECIFIC HAZARD

RATING EXPLANATION GUIDE

HEALTH		FLAMMABLE		REACTIVE	
Recommended Protection		**Susceptibility to Burning**		**Susceptibility to Energy Release**	
4	Full protection suit & breathing apparatus	4	Very flammable	4	May detonate under normal conditions
3	Full protection suit & breathing apparatus	3	Ignites under normal temperature conditions	3	May detonate with shock or heat
2	Breathing apparatus with full face mask	2	Ignites with moderate heating	2	Violent chemical change but does not detonate
1	Breathing apparatus may be worn	1	Ignites when preheated	1	Not stable if heated use precautions
0	No precautions	0	Will not ignite	0	Normally stable

Figure 4.9-5 *NFPA System*

When equipment is removed from service, dismantled and placed on a truck, an MSDS for each chemical trace left in the system must accompany the truck driver. The law requires that the equipment be flushed, cleared and tagged before it can be placed on the truck. If a hazardous chemical has been used in the system, a red tag stating, "DANGER—THIS EQUIPMENT HAS CONTAINED A HAZARDOUS MATERIAL" must be attached. Low level radioactive material falls under the plant's radioactive safety officer's responsibility and all questions should be deferred to this office.

SUMMARY

The purpose of the Hazard Communication Standard is to ensure that the hazards associated with the handling, transport, and storage of chemicals in a plant are evaluated and transmitted to affected personnel. According to the Hazard Communication Standard, chemical manufacturers and employers are required to analyze and assess the hazards associated with the chemical, develop written procedures for evaluating chemicals, document the hazard, develop an MSDS and warning labels, disseminate the information to affected individuals, and label, tag, and attach warning documentation to chemicals leaving the workplace.

Employers are responsible for:

- the development of a written Hazard Communication Program
- writing a hazardous chemical inventory list and associated MSDS
- providing training to employees about the hazards of the chemicals they will be working with
- teaching employees how to read an MSDS
- teaching employees how to select and use personal protective equipment
- using one of the three standard labeling systems: DOT, HMIS, or NFPA

Physical hazards associated with chemicals include:

- a combustible liquid
- a compressed gas
- explosive
- flammable gas
- flammable liquid
- an organic peroxide
- an oxidizer
- pyrophoric
- unstable
- water reactive

Health hazards associated with chemicals include:

- carcinogens
- mutagen
- teratogen
- reproductive toxin
- asphyxiation
- anesthetic
- neurotoxic
- allergic response
- irritants
- sensitizers
- corrosives
- toxic
- highly toxic
- target organ effects

REVIEW QUESTIONS

1. How do hazardous chemicals enter the human body?
2. Toxicology is defined as:
3. A hazardous chemical is defined as:
4. A health hazard is described as:
5. Describe a physical hazard.
6. List the typical personal protective equipment worn by a process technician.
7. A fundamental principle of the chemical hazard communication program is:
8. Define a physical hazard.
9. Chemical manufacturers are required by the HAZCOM standard to:
10. Employers are required by the HAZCOM standard to:
11. A chemical hazard communication program is composed of:
12. The development of the MSDS is the responsibility of:
13. Describe the hazard communication standard.
14. Identify the physical properties and hazards associated with handling, storing, and transporting chemicals.
15. Describe the key elements of an MSDS.
16. Describe a hazardous chemical inventory list.
17. Explain the purpose of a written hazard communication program.
18. Identify methods used to protect process technicians from hazardous chemicals.
19. Describe toxicology and the terms associated with it.
20. Describe the following terms: HMIS, DOT, and NFPA.

Chapter 5

Respiratory Protection

Government Mandate:

Respiratory Protection 29 CFR 1910.134

OBJECTIVES

- *Explain the purpose of respiratory protection.*
- *Describe air purifying respirators.*
- *Describe air supplying respirators.*
- *Describe air purifying, half-face respirators.*
- *Describe air purifying, full-face respirators.*
- *Describe an air supplying self contained breathing apparatus (SCBA).*
- *Describe an air supplying hose line respirator.*
- *Explain how to take care of and use a respirator.*
- *Analyze and contrast the limitations of each type of respirator.*
- *Review the procedures for donning and doffing air purifying respirators.*

KEY TERMS

Air purifying respirator—mechanically filters or absorbs airborne contaminants.

Air supplying respirator—provides the user with a contaminant-free air source.

Don (a respirator)—put a respirator on.

Filter life—the estimated amount of service time an air purifying filter has.

Fit testing—a critical part of a respiratory protection program; it provides respirator training, matches respirators to a technician's face structure, identifies specific chemical hazards in assigned area, and provides information on how to locate and dispose of filters, cartridges, and respirators.

IDLH—immediately dangerous to life and health.

Improper filter—the selectivity characteristics of a respirator. Process technicians must select the correct respirator to remove a specific contaminate.

National Institute for Occupational Safety and Health (NIOSH)—responsibilities include testing and certifying protective devices and research in safety and health issues in the workplace.

Negative fit test—to perform a negative fit test, the straps on the respirator should be pulled snug to ensure a tight seal. By placing the open palm over the respirator's inlet(s) a vacuum should be created when you inhale. Hold the vacuum for ten seconds to ensure a good seal.

Overpowering—atmospheric concentrations that exceed the limitations of the respirator.

Positive fit test—is performed by exhaling and covering the exhaust ports of a respirator. If the respirator has a good seal, a positive pressure will fill the face piece.

Selectivity—the specific compound or contaminate that a respirator is designed to remove.

5.2 WRITTEN RESPIRATORY PROTECTION PROGRAMS

The Occupational Safety and Health Administration requires employers who use and issue respirators to develop a written respiratory protection program. The purpose of respiratory protection is to protect employees from environments that could be hazardous. Employees must receive proper training in respiratory protection. Typically, the industrial hygiene department administers the written respiratory protection program and coordinates employee training.

Technicians are fit-tested for each respirator they will use on their units. Fit testing is a procedure that ensures that the respirator(s) forms a good seal on the technician's face. Since the facial structure of each technician is different, it is impossible to create a "one design fits all" respirator. During fit test training, technicians learn:

- which respirators will be used in their assigned areas
- which respirators fit their faces
- how to properly don respirators
- how to select respirators
- specific hazards in their assigned units

Process technicians use two basic types of respirators: air purifying and air supplying.

5.3 SELECTING A RESPIRATOR

Process technicians select respirators for use with specific substances. Since respirators are very selective in what they will remove and the protection they will provide, a number of questions must be addressed before a respirator can be selected for a specific job.

- type of contaminate or hazard
- airborne chemicals or contaminates present
- hazards associated with the chemical(s)
- physical characteristics of the chemical(s)
- air purifying or air supplying respirator
- oxygen deficient environment
- specific procedure that will be used
- personal protective equipment required
- permits required
- first aid
- length of time to perform task(s)
- exposure time limit
- concentration of chemicals present
- Will the respirator I choose protect me from the environment I will be working in?

After the known contaminates are identified, the process technician is required to reference the MSDS for specific information.

Process technicians always should be concerned about the environment they are working in and the chemicals they are handling. For this reason, the various processes inside a plant are monitored carefully. Industrial hygienists periodically sample and monitor the conditions that exist at various units over extended periods of time. Historical data is combined with current trends to determine the most effective procedure for limiting hazards.

NIOSH-approved air purifying respirators are color coded so that filters, cartridges and respirators can be matched up easily. In most plants, the contaminate they are designed to remove

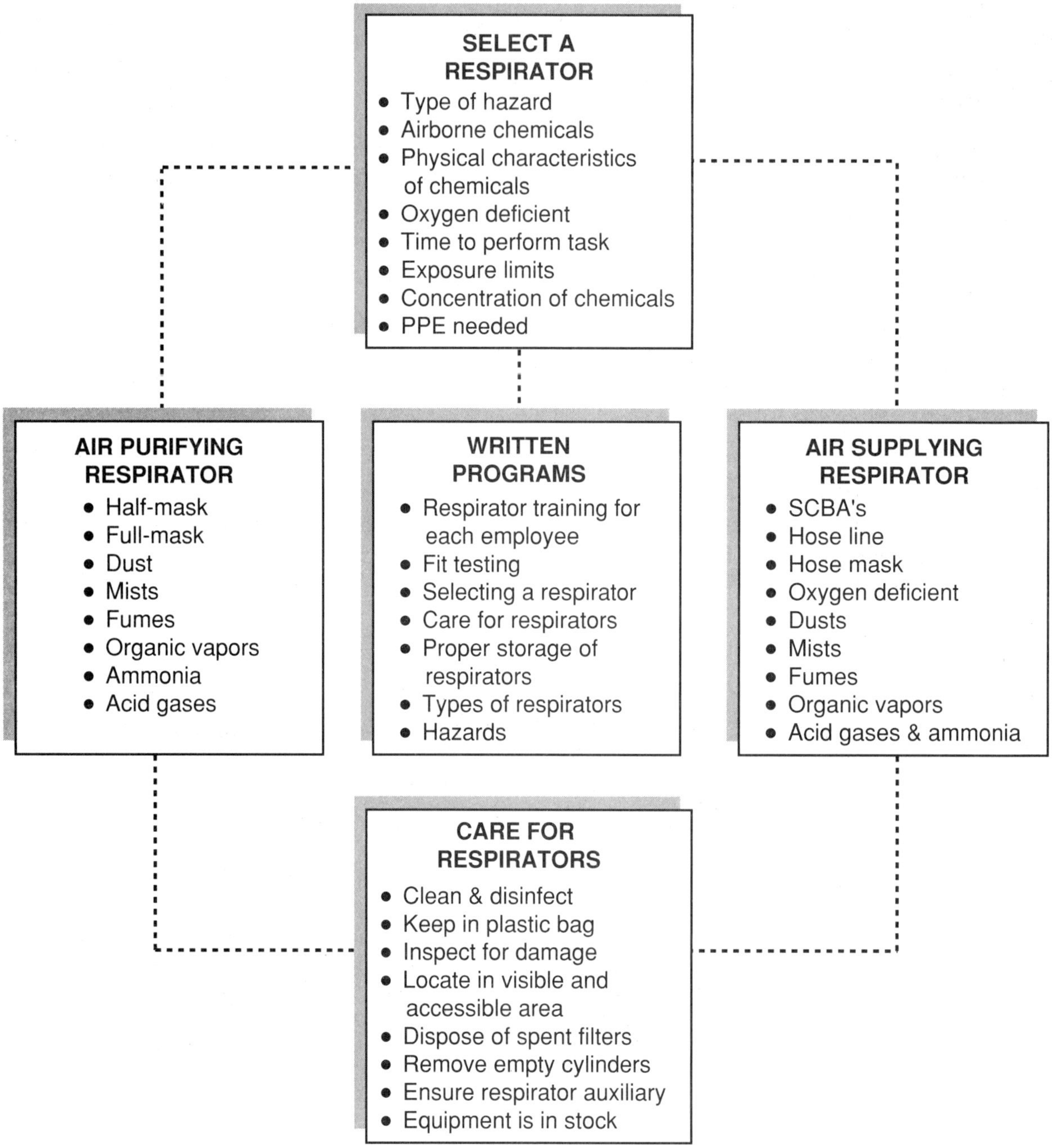

Figure 5.3-1 *Respirator Program 29 CFR 1910.134*

can be found written on the band. Air purifying respirators are designed to remove concentrations of five to fifty times the exposure limit of dust, mists, fumes, organic vapors, ammonia, and acid gases. Air supplying respirators provide protection from dust, mists, fumes, organic vapors, ammonia, acid gases, and oxygen deficient environments.

5.4 CARING FOR AND USING RESPIRATORS

Respirators should always be stored in plastic bags to keep dust and dirt from accumulating on them. OSHA inspectors will cite a company for the "improper storage of a respirator." For this reason, respirators should never be discarded carelessly or left around. Site safety inspectors as well as first and second line supervisors will notice this infraction quickly. Respirators should be located in areas where they are visible and accessible.

When a respirator is used it is the responsibility of the technician to:

- carefully clean and disinfect the device
- return the respirator to a plastic bag
- ensure that the elastomer around the face piece seal is not folded or bent
- visually inspect it for damage
- properly dispose of spent filters and cartridges

Typically, the company will supply the procedure and the cleaning materials required for this assignment. Used filters and cartridges should be properly disposed of. Check your local procedure to ensure that you are complying with your plant's requirements. If the filter is contaminated with chemicals it may fall under a special disposal plan.

Periodic training is required for those using respirators in routine job tasks. SCBA users typically train monthly on donning and using the air supplying respirator. Most companies require a technician to be able to don a SCBA in sixty seconds or less.

When a respirator is used the technician should perform a positive or negative fit test. To perform a negative fit test the straps on the respirator should be pulled snug to ensure a tight seal. By placing the open palm over the respirator's inlet(s) a vacuum should be created when you inhale. Hold the vacuum for ten seconds to ensure a positive seal. The positive fit test is performed by exhaling and covering the exhaust ports. If the respirator has a good seal a positive pressure will fill the face piece. If the respirator is damaged, it should be replaced immediately.

5.5 AIR PURIFYING RESPIRATORS

Air purifying respirators remove contaminates from the atmosphere by placing a filtering media between the contaminate and the respiratory tract of a process technician. Air is pulled through a canister, cartridge, or filter and into the respirator mask during normal respiration. The filtering media will purify the air using one of two methods: mechanically filtered or chemically absorbed.

Air purifying respirators are used on specific contaminates and compounds. Extreme care should be used when selecting a respirator. The filter life of a respirator is determined by two factors: service time and contaminate concentration. The term overpowering is used to describe atmospheric conditions that exceed the design limitations of a respirator. A respirator that makes respiration difficult or allows a small percentage of the contaminate or compound to enter into the mask should be replaced immediately.

The oxygen content of an atmosphere is another important factor that a process technician needs to consider. Air purifying respirators do not work in atmospheres with less than 19.5% oxygen content. The term IDLH is applied to oxygen deficient atmosphere that are immediately dangerous to life and health. Air purifying respirators should not be confused with air supplying respirators. Respirators have limitations that must be understood fully prior to performing a task that requires the use of a respirator.

Air purifying respirators come in two basic designs: half-face or full-face. The half-face respirator fits securely over the mouth and nose while the full-face forms a seal over the face of a process technician. Respirators come in a variety of shapes and designs.

Some examples of respirators include:

- North 7700 Half-Face Respirator—removes particulates from air; used in concentrations up to ten times the exposure limit of dust, mist, fumes, organic vapors, ammonia, and acid gases

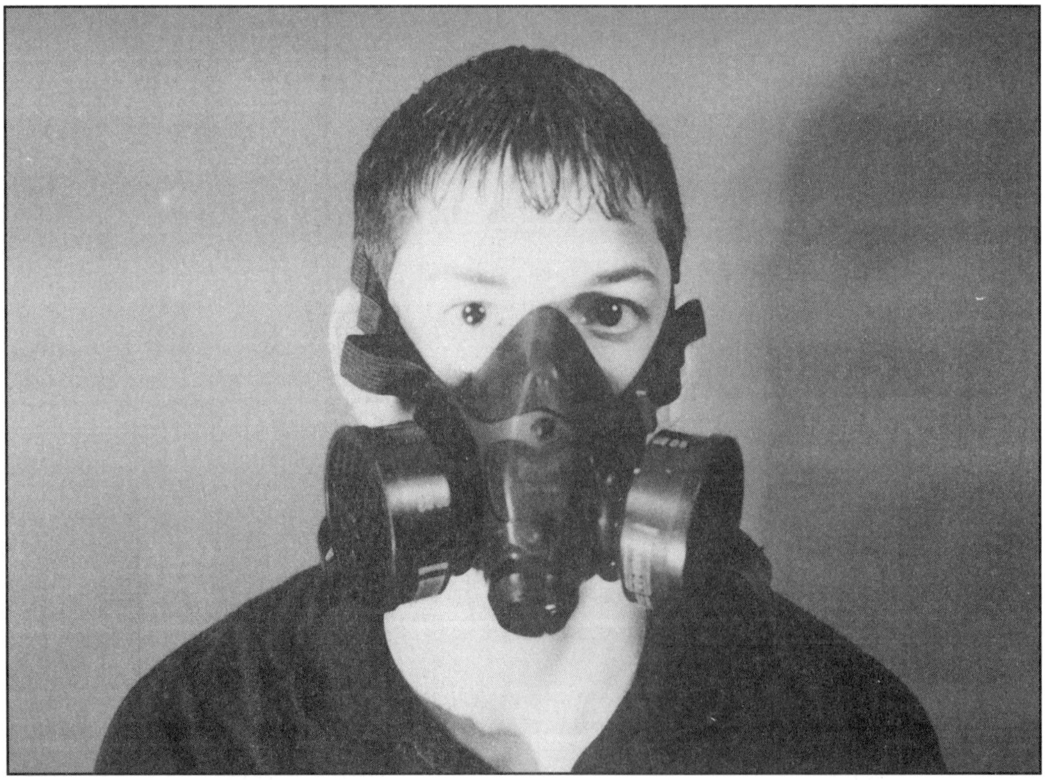

Figure 5.5-1 *Air Purifying Respirator – Half Mask*

- Scott 66 Half-Face Respirator—removes particulates from air; to be used in concentrations up to ten times the exposure limit of dust, mist, fumes, organic vapors, ammonia, and acid gases
- North 7600 Full-Face Respirator—removes particulates from air; to be used in concentrations up to fifty times the exposure limit of dust, mist, fumes, organic vapors, ammonia, and acid gases
- MSA Ultravue Full-Face piece—used for escape
- 3M 9920 Dust-Fume Mist Respirator—disposable dust mask; removes particulates from air; to be used in concentrations up to five times the exposure limit of dry materials, dusts, mists, aerosols, and welding fumes

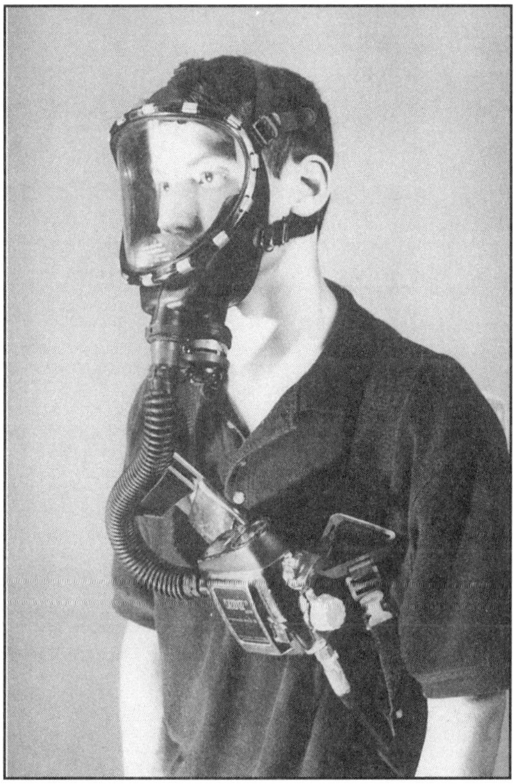

Figure 5.5-2 *Air-supplying full mask*

5.6 AIR SUPPLYING RESPIRATORS

Air supplying respirators come in two types: self-contained breathing apparatus (SCBA) and hose line respirators. Air supplying respirators are used in environments that are suspected of having toxic concentrations of chemicals or are oxygen deficient.

Self-contained Breathing Apparatus (SCBA)

Most SCBA's have full face pieces, a connecting hose, a cylinder and a harness and weigh between thirty-five and forty-five pounds. One of the first things new technicians notice after donning their air packs is the way they redistribute their center of gravity. This takes a little time to get used to and could cause an inexperienced technician to trip. Another disadvantage experienced with the SCBA is found in the face piece. On very hot or humid days, the mask fogs up and collects perspiration. Three other limitations include reduced peripheral vision, increased difficulty in getting into tight spaces or openings, and muffled, unclear communication.

SCBA's are located throughout the plant. SCBA boxes are clearly marked and should be inspected monthly. Process technicians should check the following things during an inspection:

- the five year hydrostatic test date
- physical condition of SCBA such as dents, rust, worn parts, cracks, and so on
- full charge on pressure gauge
- cylinder valve off
- red bypass valve closed
- yellow mainline regulator valve closed
- regulator cover is secure
- unthread breathing tube and ensure threading
- diaphragm
- don the SCBA and ensure proper operation

SCBA's provide a self-contained air supply that will last for about thirty minutes.

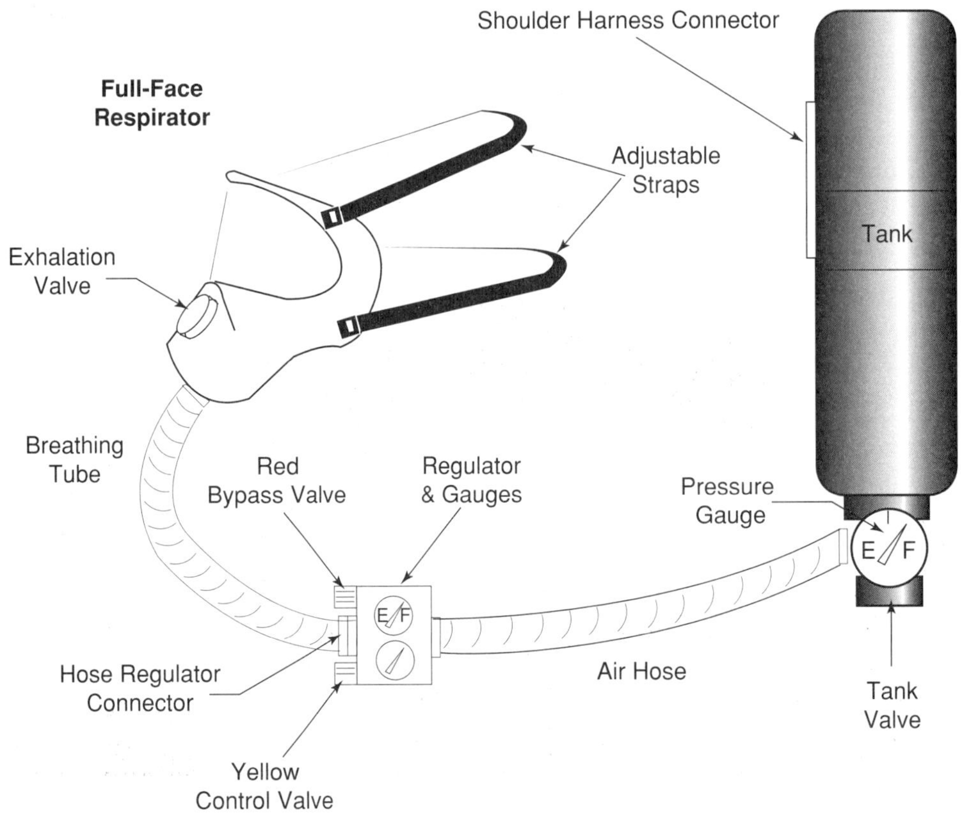

Figure 5.6-1 *SCBA*

The air supply in an SCBA will vary depending upon a number of factors:

- physical condition of the technician
- physical size of the technician
- the training and experience of the technician
- the total charge in the cylinder
- atmospheric pressure
- environmental conditions such as fear, excitement, emotional factors, and physical exertion
- carbon dioxide concentration greater than .04% in the compressed air cylinder

Hose Line Respirators

The basic components of a typical hose line respirator include a face piece, hose line, harness assembly, and air supply. Some of the more common limitations associated with hose line respirators include:

- the mask fogs up and collects perspiration
- reduced peripheral vision
- increased difficulty in getting into tight spaces or openings
- muffled, unclear communication
- weight of the hose line

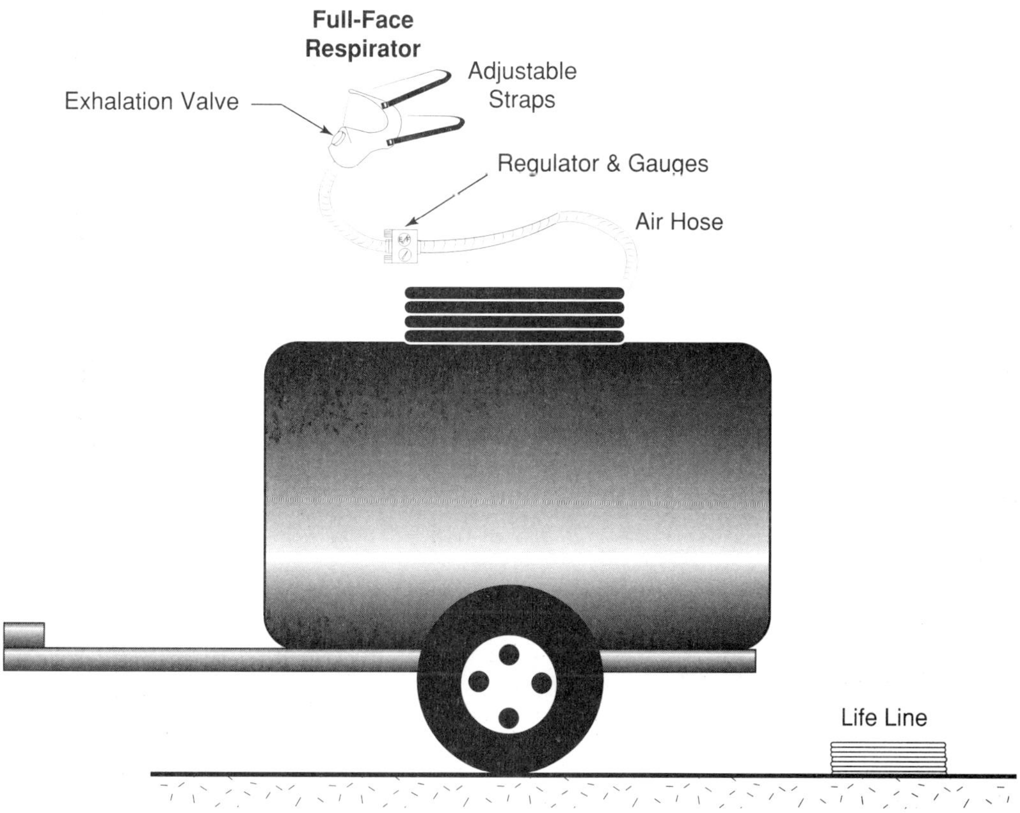

Figure 5.6-2 *Hose line Respirator*

Hose line operators are required to wear emergency escape units. These units provide five minutes or more of emergency air in order to allow the technician to escape quickly. Typically, hose line respirators are used in environments that are immediately dangerous to life and health. If air is interrupted to the face piece, the technician can use the emergency escape unit to leave the area.

Types of hose line respirators and face pieces include:

- Scott Ska-Pak—trailer mounted air supply
- Scott-o-Vista face piece—cone shaped lens and net type webbing with two straps.
- Scottoramic face piece—flat lens with five rubber straps

SUMMARY

The Occupational Safety and Health Administration requires employers who use and issue respirators to develop a written respiratory protection program. Employees must receive proper training in respiratory protection. Process technicians use two basic types of respirators: air purifying and air supplying. Air purifying respirators come in two designs: half-face and full-face. Air purifying respirators remove contaminates from the atmosphere by placing a filtering media between the contaminate and the respiratory tract of a process technician. The filtering media will purify the air using one of two methods: mechanically filtered or chemically absorbed. Air supplying respirators come in two designs: SCBA self-contained breathing apparatus and hose line respirators. These respirators are designed to be used in oxygen deficient atmospheres. Respirators are very selective in what they will remove and the protection they will provide. The following items should be considered prior to selecting a respirator:

- type of contaminate or hazard
- health hazards associated with the chemical(s)
- physical hazards of the chemical(s)
- air purifying or air supplying respirator
- oxygen deficient environment
- length of time to perform task(s)
- exposure time limit
- concentration of chemicals present

After the known contaminates are identified, the process technician is required to reference the MSDS for specific information. Respirators should always be stored in plastic bags to keep dust and dirt from accumulating on them.

When a respirator is used, it is the responsibility of the technician to:

- carefully clean and disinfect the device
- return respirator to plastic bag
- ensure that elastomer around the face piece seal is not folded or bent
- visually inspect for damage
- properly dispose of spent filters and cartridges

REVIEW QUESTIONS

1. List the two types of air purifying respirators.
2. List the different types of air supplying respirators.
3. Describe the limitations of air purifying respirators.
4. Describe the advantages of using an air supplying respirator.
5. Define IDLH.
6. Define filter life.
7. Define overpowering.
8. What are the atmospheric conditions required to use an air purifying respirator?
9. Contrast the terms selectivity and respirators.
10. What does it mean to don a respirator?
11. Describe fit testing.
12. Contrast negative and positive fit testing.

Chapter 6

Personal Protective Equipment (PPE)

Government Mandate:

Personal Protective Equipment 29 CFR 1910.133 & 135

OBJECTIVES

- *Identify the kinds of PPE used in a plant.*
- *Describe when and how to use PPE.*
- *Identify typical workplace hazards.*
- *Discuss the methods of hazard exposure prevention.*
- *Contrast emergency response and PPE.*
- *Describe the four levels of PPE.*
- *Describe the principles of hearing protection.*
- *Identify PPE outerwear worn by technicians.*
- *Explain the general limitations of PPE equipment.*
- *Contrast engineering/environmental controls and PPE.*

KEY TERMS

Engineering and environmental controls—hard technology improvements designed to make the work environment safer. Examples include sound proofing, installing guards to rotating equipment, automating product transfers, and so on.

Personal protective equipment (PPE)—used to protect a technician from hazards found in a plant.

Standard threshold shift (STS)—an audiometric test that is compared to a prior test(s) to determine a measurable hearing loss.

Totally encapsulating chemical protective suit (TECP)—the type of protective suit used to completely shield the wearer from the outside atmosphere. A TECP is used when exposure to chemicals that have a toxic effect through skin absorption, inhalation, and ingestion is expected.

6.2 PERSONAL PROTECTIVE EQUIPMENT

The human body has over nineteen square feet of surface area and inhales over 3,000 gallons of air per day. Since chemical exposure comes through inhalation, ingestion, and skin contact, protective measures need to be in place. PPE provides an effective means for protecting technicians from hazardous situations. Engineering and environmental controls provide another layer of protection. PPE is designed by nature to be used in environments that place technicians in contact with a hazardous situation. The primary purpose of PPE is to prevent exposure to hazards when engineering or environmental controls cannot be used.

Most of the hazards found in the chemical processing industry have been assessed and have PPE that will provide some degree of protection. Modern hazard management controls workplace hazards with engineering and/or environmental systems. This limits or reduces the amount of exposure a technician may have to a hazardous condition.

Outerwear

Typical outerwear worn by process technicians include:

- Safety hats—provide protection to the head; may become brittle over time; must be replaced when time stamp is exceeded; hard hats should neither be altered nor have holes drilled in them, no webbing should be removed, and they shouldn't be painted.
- Safety glasses—provide protection to the eye from front impact; safety glasses with side shields provide minimal protection from windblown objects.

Note: Most of the injuries reported in a plant are related to the eye. Safety glasses do not provide an air tight fit around the eyes and airborne contaminates can still find their way in. Caution should be exercised when working with chemicals or in dusty environments. Another limitation that clear safety glasses have is the lack of protection they provide from ultraviolet or infrared light sources. Example: welding or cutting.

- Fire retardant clothing—protects a technician for a limited amount of time from a flash fire or heat source. The clothing is specially treated and will resist bursting into flames upon contact but will not protect a technician during sustained periods of exposure. The material does not breath like cotton and is uncomfortable.
- Safety shoes—provide protection from falling or rolling objects. A typical safety shoe policy requires all technicians to wear safety shoes that have all-leather uppers and steel toes. Canvas type shoes, thongs, sandals, open-toed or high heels are not allowed in the plant. Some of the limitations of a safety shoe include chemical contamination through absorption and chemicals that enter through the top of the shoe.
- Hearing protection—protects technicians from noise that can permanently damage the inner ear. Ear plugs and ear muffs are commonly used to keep exposures below eighty-two decibels for twelve hours or eighty-five for eight hours. Ear plugs and muffs are only as effective as the technician's knowledge about correctly using them. If they are installed incorrectly they will not provide enough protection.
- Gloves—provide protection from chemicals, abrasives, or cutting surfaces. Gloves are made of a variety of compositions and are designed to work in specific situations. Some chemicals will penetrate gloves with certain compositions. Glove protection can be split into two categories: chemical and physical. Chemically resistant gloves provide protection from specific chemicals. For example, nitrile gloves are approved to work with benzene, MTBE, toluene and xylene while black plastic PVC gloves can be used with oils and other liquids. Physical hazards to the hand can be avoided by wearing leather or cotton gloves. These types of gloves provide protection from sharp or abrasive surfaces.
- Face shield—prevents chemical splashes and contaminates from getting into the eyes. The face shield provides greater coverage area than monogoggles. They will not protect the technician from airborne chemicals, fumes, or ultraviolet or infrared light sources. Face shields do not provide an air tight fit around the eyes so airborne contaminates can still find their way into the eye. Caution should be exercised when working with chemicals or in dusty environments.
- Chemical monogoggles—prevent chemical splashes and contaminates from getting into the eyes. Monogoggles will not protect the technician from airborne chemicals, fumes, or ultraviolet or infrared light sources.
- Slicker suits—provide protection from rain and inclement weather. Slicker suit material is not typically fire resistant. They can be worn when there is a potential for exposure to an acid, caustic, or chemical.
- Radio—provides emergency and operational information to other technicians and supervisors.

- Respirators—provide protection from chemicals and airborne contaminates. There are two types; air purifying and air supplying. Respirators are chemically and contaminate selective and the right one must be selected for a specific job.
- Chemical suits—provide protection from specific chemicals.
- Totally encapsulating chemical protective suits—come in a variety of shapes and designs. They are individually unique and must be used with specific substances and situations. TECP suits provide the maximum amount of protection when engineering controls are not sufficient. This type of suit has a self-contained air system and can be used for a limited time.

Figure 6.2-1 *Personal Protective Equipment*

6.3 HAZARDS IN THE WORKPLACE

Hazards in the chemical processing industry can be classified as unplanned situations that can injure or kill a process technician. Examples of hazards include burning, exploding, cutting, electrocution, impaling, tripping, falling, slipping, crushing, loud noises, or exposure to a chemical. Exposure to chemical hazards may expose a technician to cancer-causing chemicals, toxic chemicals, asphixiants, chemical burns, or irritants.

Figure 6.3-1 *Hazards in the Workplace*

Figure 6.3-2 *Hazards in the Workplace*

6.4 HAZARDS PREVENTION

Some of the methods used by industrial manufacturers to prevent exposure include engineering or environmental controls, personal protective equipment, safe work practices, good housekeeping, and training. Engineering controls include reduction of noise through the use of noise- or sound-absorbing material and new technology that automates the processes. Safe work practices include written procedures and effective employee training. Engineering controls and safe work practices are the preferred method of hazard prevention by OSHA. When these two methods cannot be used, PPE should be used to prevent exposure.

LEVEL A

- Highest Level of PPE
- Requires Totally Encapsulating Chemical Protective Suit (TECP)
- Provides Maximum Protection

↓

LEVEL B

- Chemical Exposures Not Considered Toxic Unless Absorbed through Skin
- Requires Non-Airtight Chemical Protective Suit with Taped Sleeves

↓

LEVEL C

- Hazard will not Adversely Affect Skin
- Hazards must be Identified, PPM tested, Correct Respiratory Equipment Identified, PPE Listed

↓

LEVEL D

- Minimal Amount of Protection
- Work Uniform Satisfies Level D Standards

Figure 6.5-1 *Four Levels of PPE*

6.5 EMERGENCY RESPONSE— 4 LEVELS OF PPE

According to the EPA and OSHA, emergency response has four levels of PPE. Level A requires the highest level of PPE protection by requiring a technician to don a TECP suit. The TECP suit is used in situations where inhalation and skin absorption of hazardous chemicals are possible. TECP suits provide a limited, positive air supply to the technician; they come in a variety of designs and material compositions. The chemical suit is designed to work with specific chemicals and situations. Level B deals with chemical exposures that are not considered to be extremely toxic unless they are absorbed through the skin. In this case, a non-air tight chemical protective suit may be worn. Typically, the openings on a non-air tight chemical suit are taped to limit exposure. Level C is used when the hazard is determined to not adversely affect exposed skin. Before level C protection is chosen, the following steps must be completed:

- the chemical hazard or contaminate must be identified
- the concentration (parts per million) must be identified
- correct respiratory protection must be identified
- a list of suggested PPE must be provided; for example, chemical resistant gloves.

Level D provides the minimal amount of protection to a process technician. This level takes into account that engineering controls are in place and that the technician is safe to work in this environment. Level D protection is determined by individual companies since the standard PPE is the work uniform. This may include fire retardant clothing, safety glasses, hard hat, steel-toed shoes, hearing protection, radio, and so on.

SUMMARY

Some of the methods used by industrial manufacturers to prevent exposure include engineering or environmental controls, PPE, safe work practices, good housekeeping, and training. Engineering controls and safe work practices are the preferred method of hazard prevention by OSHA. When these two methods cannot be used, PPE should be used to prevent exposure. Chemical exposure comes through inhalation, ingestion, injection and absorption or skin contact. PPE provides an effective means for protecting technicians from hazardous situations. Engineering and environmental controls provide another layer of protection. PPE is used in environments that place technicians in contact with hazardous situations. Typical outerwear worn by process technicians includes:

- safety hats
- safety glasses
- fire retardant clothing
- safety shoes
- hearing protection
- gloves
- face shields
- chemical monogoggles
- slicker suits
- radio
- respirators
- chemical suits

Emergency response has four levels of PPE: Level A requires the highest level of PPE protection, Level B deals with chemical exposures that are not considered to be extremely toxic unless they are absorbed through the skin, Level C is used when the hazard is determined to not adversely affect exposed skin, and Level D provides the minimal amount of protection and is determined to be the work uniform.

REVIEW QUESTIONS

1. Identify types of PPE used in a plant.
2. Describe when and how to use PPE.
3. Identify typical workplace hazards.
4. Discuss the methods of hazard exposure prevention.
5. Contrast emergency response and PPE.
6. Describe the four levels of PPE.
7. Describe the principles of hearing protection.
8. List PPE outerwear worn by technicians.
9. Explain the general limitations of PPE equipment.
10. Contrast engineering/environmental controls and PPE.

Chapter 7

The Permit System

OBJECTIVES

- *Describe the different types of permit systems found in the chemical processing industry.*
- *Describe hot work permits 29 CFR 1910.119*
- *Describe energy isolation and lockout/tagout 29 CFR 1910.147.*
- *Analyze the confined space entry procedure 29 CFR 1910.146.*
- *Review opening and blinding permits.*
- *Explain how to utilize a permit to enter.*
- *Describe an energy isolation permit.*
- *Recognize an unplugging permit.*
- *Describe cold work permits.*

KEY TERMS

Blinding—the installation of slip blinds between pipe flanges. Blinding isolates a process stream and allows a mechanical craftsperson to work on a piece of equipment safely.

Blind tracking—keeps a record of all blind installations. The unit blind book prevents unit start-up upsets and improves efficiency.

Electrical lockout log—includes the name of the person locking out the equipment, the date, time, and location. This same information is recorded when the lock is removed.

Enclosed space—any space that has restricted entry.

Mechanical permit acceptor—the person who accepts and returns a work permit.

Mechanical person—the person who performs the work.

Oxygen deficiency—atmospheres with less than 20% oxygen.

Process representative—the first line supervisor or designated representative who owns the equipment.

Routine hot work area—mechanical or fabrication shops.

Standby—a technician who is certified and trained to support and warn technicians who have entered a confined space.

7.2 TYPES OF PERMITS

Describing the different permitting systems used by the chemical processing industry is a difficult process. Variations between plants are enormous with only a few common permit systems. Each plant has it's own permitting systems that address routine work and maintenance, hot work, confined space entry, isolation of hazardous energy or lockout/tagout, opening piping, blinding lines, removing chemical waste, and unplugging lines.

NOTE: There are a number of permit systems that will not be discussed in this chapter. Refer to your plant's specific program for this training.

There are three permit systems that are common within the industry: the control of hazardous energy (lockout/tagout), confined space entry, and hot work. These three permitting systems are common within the CPI because they are government mandated programs. Examples of a permitting system could include:

- Hot work permit—any maintenance procedure that produces a spark, excessive heat, or requires welding or burning.
- Energy isolation procedure, lockout/tagout—isolates potentially hazardous forms of energy, electricity, pressurized gases and liquids, gravity, and spring tension. The standard is also designed to shutdown a piece of equipment at the local start/stop switch, turn the main breaker off, attach lockout adapter and process padlock, try to start the equipment, and tagout and record in lockout logbook.
- Confined space entry, permit to enter—designed to protect employees from oxygen deficient atmospheres, hazardous conditions, power driven equipment, and toxic and flammable materials.
- Opening/blinding permit—removing blinds, installing blinds, or opening vessels, lines and equipment.
- Routine maintenance permit—general maintenance and mechanical work.
- Unplugging permit—barricades area, clears lines for unplugging, informs personnel, issues opening blinding permit, and issues unplugging permit.

7.3 HOT WORK PERMITS 29 CFR 1910.119

The purpose of a hot work permit is to protect personnel and equipment from explosions and fires that might occur from hot work performed in an operational area. Hot work is defined as any maintenance procedure that produces a spark, excessive heat, or requires welding or burning. Examples of hot work include portable grinders, open fires, welding equipment, energized electrical circuits, internal combustion engines, electric motors, hot plates, coriolis meter, turbine meter, concrete busters, soldering irons, dry sandblasting, and so on. The hot work permit has multiple layers of protection. One large chemical processing company involves the following people during the issue of a hot work permit:

- process technician
- first line supervisor
- person performing the work
- mechanical supervisor
- safety permit inspector
- standby

The process technician is responsible for securing the process to be worked on. If the lines are full of hydrocarbons or toxic materials they must be drained, inerted (explosive or hazardous vapors removed) and blinded. All sewers and manholes must be covered and the appropriate fire fighting equipment placed at the work site. Before calling the safety inspector, the area should be cleaned up, if needed. Good housekeeping is an essential element of the hot work permit procedure. If the hot work will be conducted inside a confined space, then all of the procedures for confined space entry come into play and must be completed before the hot work permit can be filled out.

BIGG CHEMICAL AMERICAS - CHEMICAL PLANT

HOT WORK PERMIT

(WELDING . BURNING)

LOCATION ______

DATE ISSUED ______
PERMIT EXPIRES ______
SAFETY PERMIT INSPECTOR ______
FIRE EQUIPMENT ON HAND READY FOR USE:
WATER CO_2 DRY CHEMICAL

SPECIAL PRECAUTIONS: ______

	DATE	TIME
BASIC EQUIPMENT OWNER		AM PM
MECHANICAL SUPERVISOR		AM PM
BASIC EQUIPMENT OWNER		AM PM
MECHANICAL SUPERVISOR		AM PM
BASIC EQUIPMENT OWNER		AM PM
MECHANICAL SUPERVISOR		AM PM
BASIC EQUIPMENT OWNER		AM PM
MECHANICAL SUPERVISOR		AM PM

Figure 7.3-1 *Hot Work Permit Example*

The hot work permit must be filled out and ready for the appropriate signatures before the inspector is called. When the system is ready, the mechanical supervisor and safety inspector will show up to inspect the area. Chemical concentrations and potential hazards are assessed. The need for a standby will be determined by the mechanical supervisor and the basic equipment operator. If everything is in order, the safety inspector will sign the card and the work can be started.

An example of typical roles and responsibilities:

1. The process technician
    - inspects area and ensures housekeeping
    - blinds, isolates, and clears equipment, vessels, tanks and piping
    - immobilizes power driven equipment (lockout/tagout procedure)
    - determines PPE required
    - fills out the permit and posts it at the job site

2. The first line supervisor
   - delegates responsibilities to the process technician and ensures that all established procedures are completed

3. The person performing the work
   - inspects the job site
   - gathers information from process representative and mechanical supervisor about potential hazards, special procedures, or conditions
   - selects and dons appropriate safety equipment
   - performs work (other permits may be required)

4. The mechanical supervisor
   - inspects area and ensures that it is ready for safety inspector
   - ensures that equipment, vessels, and piping are cleared
   - ensures that safety equipment is located near job site
   - reviews procedure with person performing the work
   - confirms PPE required
   - signs permit

5. The safety permit inspector
   - inspects area and ensures that it is safe
   - performs gas test and determines oxygen level
   - ensures that equipment, vessels, and piping are cleared
   - confirms PPE required
   - signs permit and sets time limit

6. The standby
   - ensures that the person performing the work is safe
   - wears the PPE required to perform the job
   - warns the person performing the work if a hazardous condition develops (may use a Klaxon horn, sign language, or radio)

Note: The standby does not go into the confined space until help arrives.

According to Process Safety Management Standard 29 CFR 1910.119, "Hot Work" employers are required to issue a permit for hot work operations. The standard defines hot work as involving welding, cutting, brazing, or similar spark-producing operations. The permit must indicate:

- that fire prevention and protection measures were in place before the hot work was initiated
- the date(s) the permit is approved for
- the location and equipment used, item where hot work is performed
- that the fire watch was posted and in place during procedure and thirty minutes after work is complete

The permit must be displayed at the work site until the hot work operation is complete.

7.4 CONFINED SPACE ENTRY

The Permit Required Confined Spaces for General Industry is part of the OSHA 29 CFR 1910.146 standard. The chemical processing industry has a variety of areas where confined space entry routinely takes place. Confined space awareness includes:

- define confined space entry
- describe the hazards associated with confined space entry
- who is involved in the confined space procedure
    - process representative
    - mechanical supervisor
    - person performing the work
    - standby
- hazards
- describe testing and monitoring equipment, ventilation equipment, radio equipment, and retrieval equipment

Confined spaces have restricted or limited access. Entry occurs when the technician's body breaks the plane of the opening. A confined space is defined as a space large enough for a person to enter, has restricted entry, and is not designed for continuous occupancy. This includes excavations, sewers, pits, reactors, boilers, furnaces, distillation columns, strippers, absorbers, vessels, tanks, silos, blenders, drums, piping, pumps, compressors, heat exchangers, extruders, or any space that restricts entry or allows the head of the worker to go below the top of the confined space.

A confined space may contain a hazardous atmosphere, chemical, or asphyxiate, or potential health hazard. Inside a refinery or chemical plant, a variety of confined spaces, vessels, piping, excavations, and power driven equipment exist. The permit to enter is designed to protect employees from oxygen deficient atmospheres, hazardous conditions, power driven equipment, and toxic and flammable materials. The permit to enter must be completed before any enclosed space can be entered. The permit covers confined space entry only. Additional permits are required for cold work, hot work, opening/blinding, or energy isolation inside the enclosed space. Confined space entry is a complex procedure that requires a number of individuals to work together as a team. Several large industrial manufacturers utilize the team approach when dealing with a confined space entry.

Upon initial assignment to the unit, process technicians are given confined space awareness training. This training identifies what a confined space is and specific hazards it presents. Hazards from a confined space include asphyxiation, engulfment, heat stress, electrocution, moving equipment, getting wedged in a narrow space, and falling. These hazards can be removed when the steps for confined space entry are followed.

Employers are required to develop a written confined space entry program, identify all permit places in their plant, post warning signs, prevent unauthorized entry into confined spaces, develop procedures, provide training in all of these areas, and use PPE. Process technicians

CONFINED SPACE ENTRY PERMIT 012345

Date and Time Issued: ____________________
Job Site/Space I.D.: ____________________
Equipment to be worked on: ____________________

Date and Time Expires: ____________________
Job Supervisor ____________________
Work to be performed: ____________________

Stand-by personnel: ____________________ ____________________ ____________________

1. Atmospheric Checks: Time ________
   Oxygen ________ %
   Explosive ________ % L.F.L.
   Toxic ________ PPM

2. Tester's signature:
3. Source isolation (No Entry): N/A Yes No
   Pumps or lines blinded, ( ) ( ) ( )
   disconnected, or blocked ( ) ( ) ( )

4. Ventillation Modification: N/A Yes No
   ( ) ( ) ( )
   ( ) ( ) ( )

5. Atmospheric check after isolation and ventillation:
   Oxygen ________ %
   Explosive ________ % L.F.L.
   Toxic ________ PPM
   Time ________
   Tester's signature ____________________

6. Communication procedures: ____________________
   ____

7. Rescue procedures: ____________________
   ____
   ____

8. Entry, standby, and back up persons:
   (Successfully completed required training?) Yes No
   Is it current? ( ) ( )

9. Equipment: N/A Yes No

	N/A	Yes	No
Direct reading gas monitor tested	( )	( )	( )
Safety harness & life for entry and standby persons	( )	( )	( )
Hoisting equipment	( )	( )	( )
Powered communications	( )	( )	( )
SCBA's for entry & standby persons	( )	( )	( )
Protective clothing	( )	( )	( )
All electric equipment listed Class 1, Div. 1 Group D and non-sparking tools	( )	( )	( )

10. Periodic atmospheric tests:
   Oxygen ________ % Time ________
   Oxygen ________ % Time ________
   Explosive ________ % Time ________
   Explosive ________ % Time ________
   Toxic ________ % Time ________
   Toxic ________ % Time ________

Permit prepared by: ____________________ (supervisor)

Approved by: ____________________ (Unit supervisor)

Reviewed by : ____________________ (Cs Operations Personnel)

Figure 7.4-1 *Confined Space Entry Permit*

should be aware that conditions within a confined space can change and should be monitored carefully. Welding or cutting can expose hazardous chemicals that have been absorbed by the metal, use up oxygen, and concentrate noxious fumes. Contractors may also accidentally mix chemicals inside a confined space that they have brought in for specific work. It is during this time period that the role of the standby becomes critical; the standby is required to monitor work being performed in the confined space.

A typical confined space entry permit should include:

- work to be performed
- location
- date(s) and authorization period
- atmospheric testing results and entry conditions
- entrant's name
- entry supervisor and standby names
- hazards and how to control them
- isolation and control methods
- communication procedures
- other active permits (hot work)
- rescue and emergency procedures
- equipment to be used

Preparation for Confined Space Entry

To prepare a confined space for entry, the process technician should follow the plant's procedure carefully. Since each facility is unique, each task should be given careful consideration. Typically, the process operator will secure the vessel or area that is being isolated. For example, all pipes entering and exiting the confined space should be isolated or blinded. This procedure will stop chemicals from being admitted accidentally into the confined space during work activities. Any chemical remaining in the system must be removed prior to the confined space entry taking place. This procedure may require water washing, purging, or inerting. All electrical equipment in the confined space should be de-energized, locked out, and tagged prior to entry. Adequate ventilation should be provided to the confined area. This procedure might require the use of fans or blowers and back-up fans and blowers should be available in case of failure. Operators should post signs, erect barricades, and pull barricade tape during the preparation phase. Housekeeping is also very important prior to having work performed in your unit. Once the area is clear to be worked on, the atmosphere will be tested. Oxygen concentrations must be between 19.5 and 23.5%. Flammable gas concentrations must be less than 10% of their lower flammable limit. Airborne contaminates must fall into correct guidelines or work cannot be performed. The safety inspector will require that all emergency equipment and communication procedures be in place.

During the confined space entry, the process technician will work closely with the people performing the work. Contract technicians on the entry team will not be aware of all the hazards in the confined space unless you document them and tell the technicians about them. The permit must be displayed near the job at all times and removed after the job is completed. Once the work on the confined space is initiated the process technician works with the confined space entry team.

The Confined Space Entry Team and the Process Technician
The confined space entry team is composed of an entry supervisor, a standby or attendant, and an entrant. Each member of this team will have specific training prior to performing confined space work. The entry supervisor will ensure that conditions are safe prior to admitting his or her team. During the operation, the entry supervisor frequently will monitor work conditions. The standby is required to observe conditions, be aware of hazards, recognize exposure to hazards, keep unauthorized people out, stay in constant contact with entrant, order workers out of the area if a hazard is present, coordinate rescues, and remain outside the confined space until relieved. The entrant should recognize the early signs of exposure to a hazard. For example, oxygen deficiency can cause breathing difficulty, loss of muscle control, ringing in the ears, feelings of well-being, mental confusion, and death. Entrants should be familiar with PPE and its limitations. The entrant should remove his or herself from the confined space as soon as one of these situations is detected.

7.5 CONTROL OF HAZARDOUS ENERGY (LOCKOUT/ TAGOUT) 29 CFR 1910.147

The purpose of the hazardous energy standard is to protect employees from the hazards associated with the accidental release of uncontrolled energy. The lockout/tagout procedure is a standard designed to isolate a piece of equipment from its energy source. OSHA requires employers to have a written energy isolation program and to provide training to new employees upon initial assignment and every two years thereafter. Equipment modifications and new unit start-ups require the existing workforce to have additional lockout/tagout training.

The chemical processing industry harnesses energy from seven basic forms: electrical, pneumatic, hydraulic, compressed gases, liquids, gravity, and spring tension. The two most common classifications of energy are kinetic and potential. Kinetic energy is closely associated with movement while potential energy is related to stored energy. Every year, the chemical processing industry records severe injuries because process technicians fail to disconnect electrical equipment, dissipate residual energy, stop the equipment, restart the equipment accidentally, or fail to clear the area before equipment start-up.

OSHA has established a six step procedure for locking out a piece of equipment. The first step is the preparation for shutdown phase. During this phase, the type of energy being isolated must be identified and the specific hazards controlled. Phase two involves shutting down the equipment. Phase three is the isolation step which involves the closing of valves, shutting down main disconnects and circuit breakers, and disconnecting pneumatic, electric, hydraulic and compressed gas and liquid lines. Phase four is the application of lockout/tagout devices to breakers and disconnect switches, valves, and energy isolating devices. Phase five is directed at the control of stored energy. Pressure must be relieved, grounding cables connected, elevated equipment supported, and moving parts stopped. Phase six is the verification step. The term lock-tag-try is applied when the electrically disconnected equipment is checked by attempting to start the equipment at the local start-stop switch. If the procedure has been performed correctly the equipment will not start.

The six steps of the procedure:

1. preparation for shutdown
2. shut down the equipment
3. isolation of equipment—closing valves, shutdown main disconnects and circuit breakers, disconnect pneumatic, electric, hydraulic and compressed gas and liquid lines
4. apply lockout/tagout devices to breakers and disconnect switches, valves, and energy isolating devices
5. control of stored energy—relieve pressure
6. verification step—lockout/tagout

All of this information should be recorded in an electrical lockout logbook.

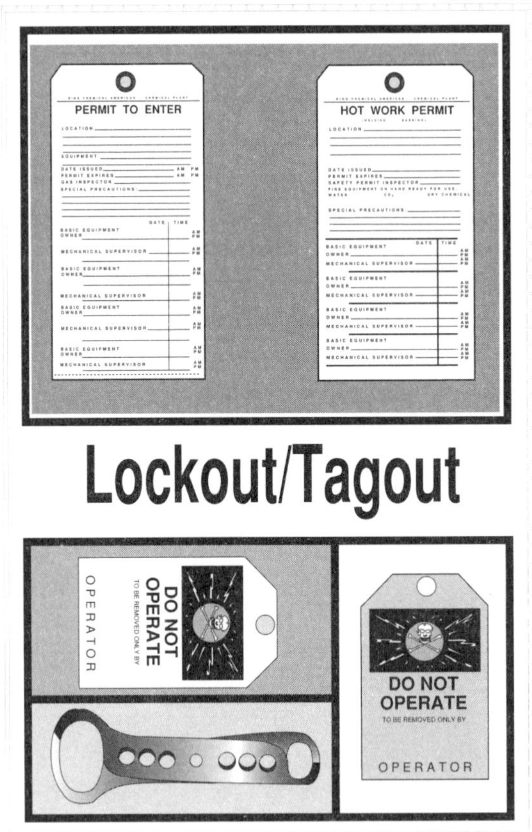

Figure 7.5-1 *Lockout/Tagout*

7.6 OPENING OR BLINDING PERMITS

The purpose of the opening or blinding permit is to limit accidents through improved communications and structured guidelines for removing blinds, installing blinds, or opening vessels, lines and equipment. Examples of work covered under opening and blinding permits include the installation and removal of blinds, the removal of hatch covers, and the opening of equipment.

To initiate an opening and blinding permit, the process technician performs a safety inspection, secures the area for work, isolates the equipment, and then fills out the opening and blinding permit. After the permit is filled out, the technician speaks with the craftperson about the work to be performed, reviews the hazards, and hangs a hard copy of the permit near the unit with the craftperson's signature. Another copy of the permit is also displayed in the control room.

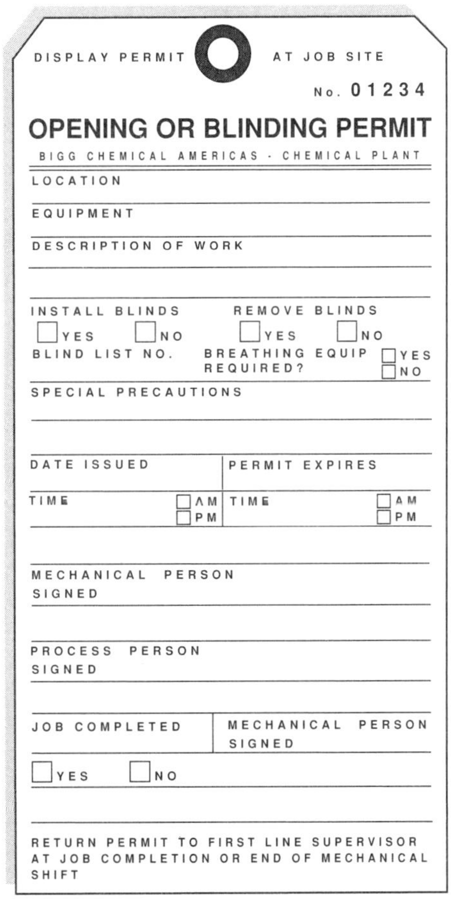

DISPLAY PERMIT AT JOB SITE

No. 01234

OPENING OR BLINDING PERMIT

BIGG CHEMICAL AMERICAS · CHEMICAL PLANT

LOCATION

EQUIPMENT

DESCRIPTION OF WORK

INSTALL BLINDS ☐ YES ☐ NO

REMOVE BLINDS ☐ YES ☐ NO

BLIND LIST NO.

BREATHING EQUIP REQUIRED? ☐ YES ☐ NO

SPECIAL PRECAUTIONS

DATE ISSUED | PERMIT EXPIRES

TIME ☐ AM ☐ PM | TIME ☐ AM ☐ PM

MECHANICAL PERSON SIGNED

PROCESS PERSON SIGNED

JOB COMPLETED ☐ YES ☐ NO | MECHANICAL PERSON SIGNED

RETURN PERMIT TO FIRST LINE SUPERVISOR AT JOB COMPLETION OR END OF MECHANICAL SHIFT

Figure 7.6-1 *Opening or Blinding Permit*

7.7 ROUTINE MAINTENANCE PERMITS

The routine maintenance permit system is designed to prevent accidents by improving communications between process technicians and the craftspersons doing the mechanical work. In order to issue the routine maintenance permit the technician is required to check out the area where the work will be performed. Safety hazards must be identified and the permit must be displayed near the work site. Another copy of the permit must also be displayed in the control room. Both the process technician and the craftsperson sign the permit.

The process technician is responsible for:

- performing safety inspection prior to issuing the permit
- filling out the routine maintenance permit
- speaking with the craftsperson about the work to be performed
- reviewing hazards
- hanging hard copy of permit near unit with craftsperson's signature on it
- posting a second copy of the permit in the control room
- extending the permit, if necessary

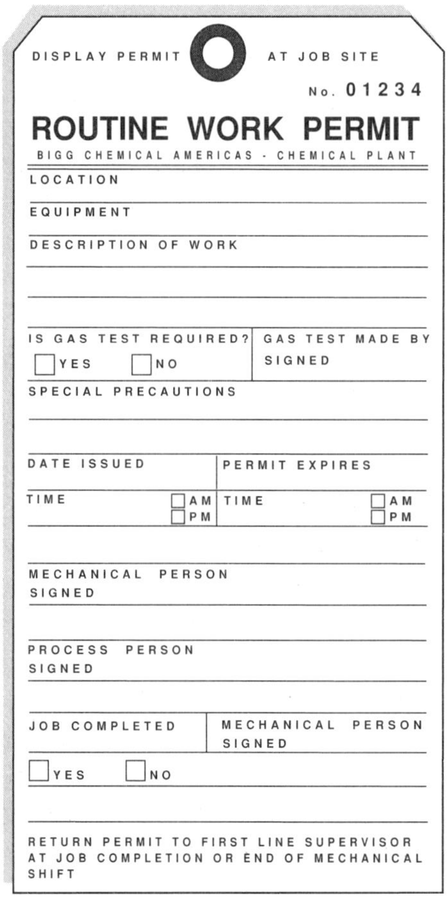

DISPLAY PERMIT AT JOB SITE

No. 01234

ROUTINE WORK PERMIT

BIGG CHEMICAL AMERICAS - CHEMICAL PLANT

LOCATION

EQUIPMENT

DESCRIPTION OF WORK

IS GAS TEST REQUIRED? ☐ YES ☐ NO

GAS TEST MADE BY SIGNED

SPECIAL PRECAUTIONS

DATE ISSUED | PERMIT EXPIRES

TIME ☐ AM ☐ PM | TIME ☐ AM ☐ PM

MECHANICAL PERSON SIGNED

PROCESS PERSON SIGNED

JOB COMPLETED ☐ YES ☐ NO

MECHANICAL PERSON SIGNED

RETURN PERMIT TO FIRST LINE SUPERVISOR AT JOB COMPLETION OR END OF MECHANICAL SHIFT

Figure 7.7-1 *Routine Maintenance Permit*

The mechanical craftsperson is responsible for:

- initiating the cold work permit
- performing safety inspections
- signing the cold work permit
- performing authorized work
- securing permit extensions
- performing housekeeping after the job is complete
- completing work, signing the card, and returning it to the process representative
- securing process approval and updating the permit board

SUMMARY

The types of permits used in the chemical processing industry include:

- routine maintenance permit—general maintenance and mechanical work that does not involve hot work or opening up a vessel
- hot work permit—any maintenance procedure that produces a spark, excessive heat, or requires welding or burning
- opening/blinding permit—removing blinds, installing blinds, or opening vessels, lines and equipment
- confined space permit to enter—protects employees from oxygen deficient atmospheres, hazardous conditions, power driven equipment, and toxic and flammable materials
- energy isolation procedure, lockout/tag-out procedure—isolates potentially hazardous forms of energy, electricity, pressurized gases and liquids, gravity, and spring tension.

The energy isolation procedure is designed to shutdown a piece of equipment at the local start/stop switch, turn the main breaker off, attach the lockout adapter and process padlock, try to start the equipment, and tagout and record in the lockout logbook.

REVIEW QUESTIONS

1. List the three permit systems used by the chemical processing industry.

2. Define hot work.

3. Explain the steps you would take to prepare a tank for confined space entry.

4. Pump 102A is running and needs to be removed for maintenance. Pump 102B is ready to be placed in service. List the steps you would go through to prepare P-102A for maintenance.

5. The maintenance supervisor informs you that at 14:00 the welder will show up to weld a small line to the suction side of pump 104A. Describe the steps you would go through to prepare for this activity.

6. A small water valve is scheduled to be replaced by maintenance. Explain the steps you would go through to prepare for this activity.

7. Heat Exchanger 701A is scheduled to be removed from service to have the tubes cleaned. This will require the inlet and outlet heads to be removed, so the tubes will be exposed. Exchanger 701B will need to be placed in service. Explain the steps you would take to prepare for this activity. List all permits and PPE needed.

Chapter 8

Fire Prevention, Protection, and Control

Government Mandate:

Fire Prevention, Protection, and Control 29 CFR 1910.157

OBJECTIVES

- ***Describe the principles of fire prevention, protection, and control.***
- ***Explain the chemistry of fire.***
- ***Describe the fire classification system.***
- ***Name and evaluate the different types of fire extinguishers.***
- ***Analyze the different fire stages.***
- ***Identify the various types of fire fighting equipment.***
- ***Explain the process for responding to a fire emergency.***
- ***Describe flammable and explosive materials.***

KEY TERMS

Conduction—occurs as the molecules that make up a solid begin to vibrate and transfer energy across their matrix.

Convection—occurs as the molecules that make up a liquid or a gas speed up as heat energy is added and move naturally from areas of hot to cold.

Fire hydrants—located throughout the plant to provide water for fire trucks.

Fire monitors—used to cool down exposed facilities and equipment and limit the spread of the fire. The fire monitor is equipped with an adjustable 500 gallon per minute nozzle that has three distinct spray patterns: fog, straight stream, and power cone.

Fire triangle—the three sides of the triangle are represented by fuel, oxygen, and a source of ignition. The removal of any one of these components will extinguish the fire.

Flame stage—flames are visible.

Heat—a form of energy that comes in a variety of forms: sensible heat, latent heat of vaporization, latent heat of fusion, latent heat of condensation and specific heat. Measured in BTUs or calories.

Heat stage—excessive heat, flame, smoke, and toxic gases.

Hose Reel—takes the hose to the fire. Most hose reels have an adjustable nozzle and approximately 100 feet of 1 1/2 inch hose.

Incipient stage—no smoke or flame, little heat, combustion begins.

Radiant heat—transfers energy through space by means of electromagnetic waves.

Smoldering stage—increased combustion, smoke, no visible flame.

Temperature—the hotness or coldness of a substance. Measured in degrees fahrenheit (F) or celsius (C).

8.2 FUNDAMENTALS OF FIRE PREVENTION, PROTECTION, AND CONTROL

The fundamentals of fire prevention, protection, and control include a variety of complex and technical standards. Most of these standards can be found in *The Handbook of Industrial Loss*

Prevention by Factory Mutual Engineering Corporation and *The Handbook of Fire Protection* by the National Fire Protection Association. A summary of these standards includes:

- analyze the physical layout of the plant or facility and surrounding structures
- ensure that fire fighting equipment is available and strategically placed
- design engineering systems that allow for the safe handling, storage, and shipment of flammable and combustible materials
- establish written procedures and emergency response plans
- develop structured safety training programs
- design and implement early fire detection systems
- prevent the outbreak and spread of fires

8.3 THE CHEMISTRY OF FIRE

Fire is a chemical reaction between oxygen, fuel, and heat. If any one of these three elements is removed, the fire will go out. Fire is defined as rapid oxidation with the evolution of light and heat. In the fire triangle, fuel must be present in vapor form, be in the flammable range, and there must be oxygen or an oxidizing agent; heat must be in the form of high environmental temperatures, hot surfaces, open flames, friction, and so on.

One of the greatest dangers to personnel and facilities in an industrial environment is from fire and explosions. Most potential hazards are engineered out; however, the nature of the work requires routine handling of highly flammable materials. Therefore, it is important to be constantly aware of:

- the ingredients of a fire
- fire prevention and control measures
- the properties of flammable and combustible materials

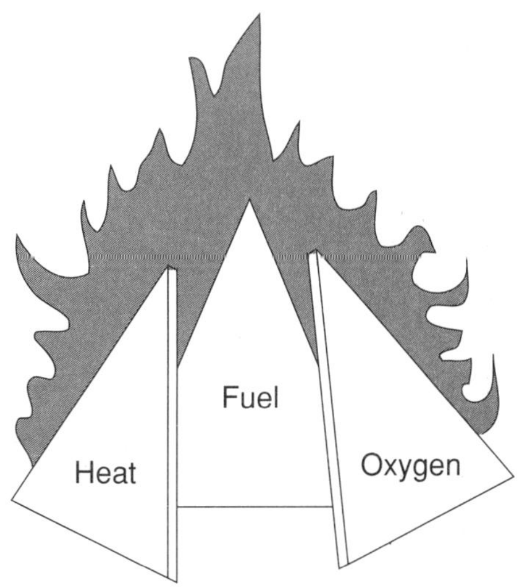

Figure 8.3-1 *The Fire Triangle*

When all of the elements of the fire triangle combine and combustion occurs, the flame both radiates and convects heat to the surrounding area. If flame impingement occurs on a metal vessel or tank, conduction carries heat from the flame to the contents of the vessel or tank. Radiant heat, conductive, and convection contribute equally once combustion takes place. The heat source in the fire triangle can be delivered by any of the three heat transfer methods.

Radiant heat provides the energy feedback from the flame to the burning material. This impacts the rate of burning. Flame spread during a fire occurs primarily through radiant heat transfer. Radiant heat may be so intense that a process technician cannot approach the flame.

8.4 FIRE STAGES

The development of a fire can be traced through four distinct stages:

1. Incipient stage: no smoke or flame, little heat, combustion begins
2. Smoldering stage: increased combustion, smoke, no visible flame
3. Flame stage: flames become visible
4. Heat stage: excessive heat, flame, smoke, and toxic gases

8.5 FIRE CLASSIFICATION SYSTEM

The fire classification system is designed to simplify the selection of fire-fighting techniques and equipment:

- Class A: involve the burning of combustible materials such as wood, paper, plastic, cloth fibers, and rubber
- Class B: involve combustible and flammable gases and liquids and grease
- Class C: categorized as electrical fires. This involves energized equipment and Classes A, B and D materials that are located near the fire
- Class D: cover combustible metals such as aluminum, magnesium, potassium, sodium, titanium, and zirconium

8.6 TYPES OF FIRE EXTINGUISHERS

Portable, hand-held fire extinguishers are carefully located around the process unit. These devices are a fundamental part of the fire prevention, protection, and control standard. Since most fires require split-second reactions, the portable fire extinguisher is an excellent first line defense. Portable fire extinguishers can be used to stop a fire before it gets out of control, provide an escape route through a large fire, and contain a fire until help or the fire team arrives.

Fire extinguishers come in a variety of shapes and designs. The selection and use of a fire extinguisher is dependent upon the type of fire being fought. Process technicians need to be familiar with the fire classification system in order to select the proper extinguisher in an emergency situation. The five most common fire extinguishers found in the chemical processing industry include:

1. CO_2 Fire Extinguisher—The carbon dioxide extinguisher is composed of a cylinder filled with compressed carbon dioxide. A dip tube penetrates the top of the cylinder and continues to the bottom. A carrying handle and operating lever are located at the top of the extinguisher. A discharge hose is attached to the dip tube on the cylinder side and an oversized discharge horn at the exit. CO_2 extinguishers are effective on Class B and C fires because they displace oxygen. Process technicians prefer CO_2 extinguishers because they are inexpensive and do not leave a chemical residue. Limitations of this type of extinguisher include poor performance in windy conditions, asphyxiation in enclosed spaces, and frostbite if it touches human tissue. A fully charged CO_2 extinguisher weighs about forty-five pounds and has an effective range of about ten feet and a twenty second lifetime.

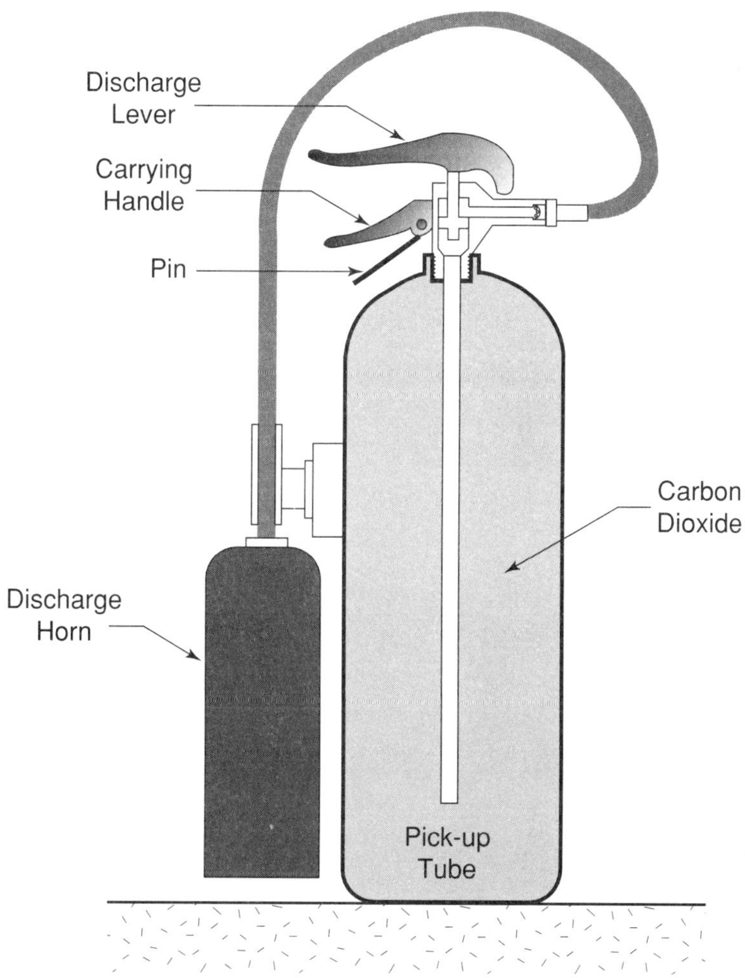

Figure 8.6-1 *Carbon Dioxide Fire Extinguisher*

2. Dry chemical fire extinguishers—This is composed of a cylinder, dip tube, pressure gauge, hose and nozzle, BC or ABC dry chemical agent, carrying handle, operating lever, locking pin, and a compressed nitrogen or carbon dioxide cartridge. The extinguisher needs to be operated in the upright position in order to pick up the extinguishing agent from the bottom of the dip tube. The dry chemical fire extinguisher can be used on Class A, B, or C fires. This type of portable extinguisher leaves a residue that can damage computer equipment and sensitive electronics. Using this device in a confined space is dusty and messy. When charging the cylinder with an 800 psi CO_2 cartridge, the extinguisher should be pointed away from the upper body and head. A fully charged dry chemical cylinder will have about 200 pounds of pressure in it. It has an effective range of thirty feet and weighs about forty pounds. It is approved for confined space work and available in 150 or 250 pound wheeled units.

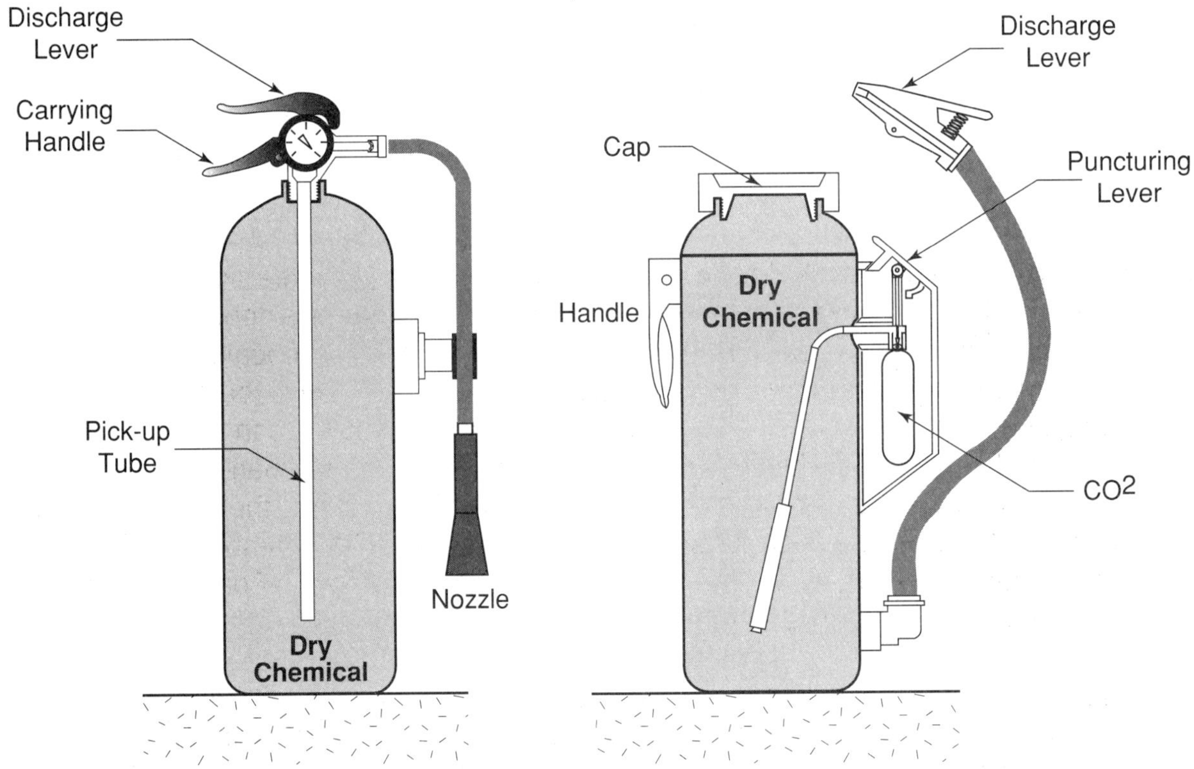

Figure 8.6-2 *Dry Chemical Fire Extinguisher (include both CO_2 cartridge and nitrogen)*

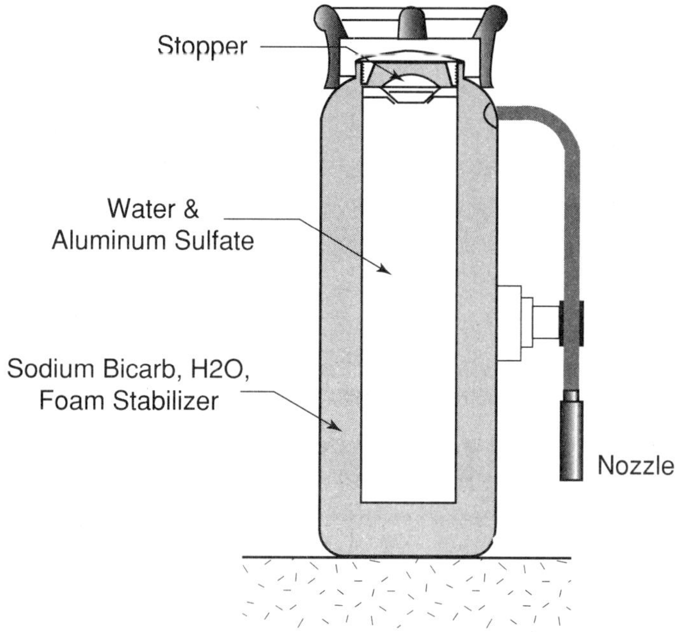

Figure 8.6-3 *Foam Fire Extinguisher*

3. Foam fire extinguishers—These are used to control flammable liquid fires. The foam forms an effective barrier between the flammable liquid and the needed oxygen for combustion. Many of the large open top tanks used to store hydrocarbons in industry are equipped with foam depression systems. Portable, hand-held, foam fire extinguishers are equipped with a cylinder that has two sections. One section contains a water and aluminum sulfate mixture while the other section contains water, foam stabilizer, and sodium bicarbonate. In order to use this type of fire extinguisher, the device needs to be inverted. This allows a lead stopper to fall out and the two chambers to combine. The combination of these two chambers forms a compressed foam. Foam extinguishers are effective on Class A and B fires.

4. Halon fire extinguisher—This is composed of a cylinder filled with high pressure Halon gas, a dip tube, hose, carrying handle and operating lever, a pressure gauge, and a locking pin. This type of extinguisher is considered to be superior to a dry chemical extinguisher because it does not leave a corrosive residue on sensitive electronics or computers. A Halon extinguisher is designed for use on Class A, B, and C fires. It has an effective range of twenty feet and a lifetime of thirty seconds. Halon extinguishers should

never be used in an enclosed space because the halon gas and fire produce chemical vapors that can damage the respiratory tract.

5. Water fire extinguishers—This is composed of a cylinder, dip tube, pressure gauge, carrying handle, locking pin, operating lever, and over fill tube. Portable, hand-held water filled fire extinguishers are designed for use on Class A fires only.

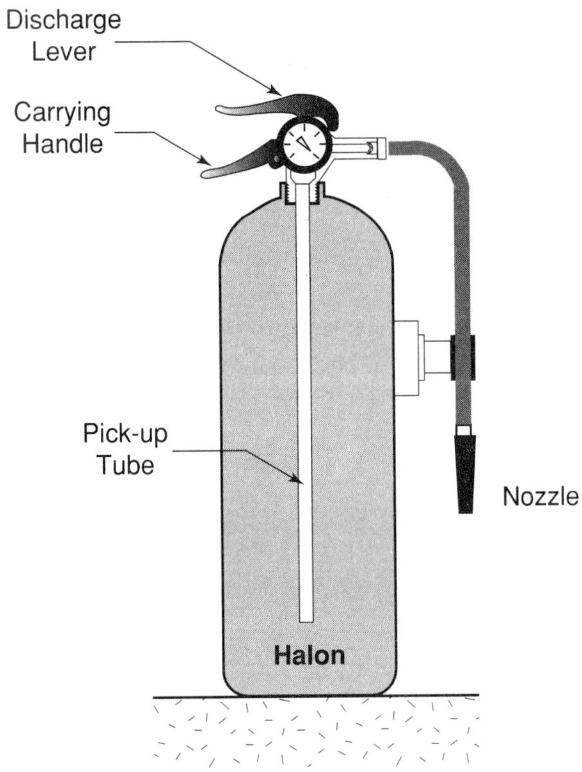

Figure 8.6-4 *Halon Fire Extinguisher*

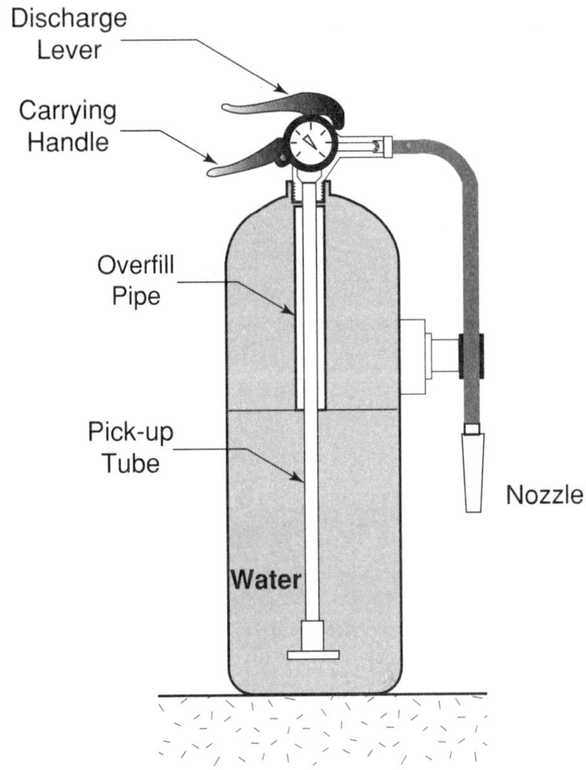

Figure 8.6-5 *Water Fire Extinguisher*

8.7 FIRE EXTINGUISHER USE

In the proper hands, a portable, hand-held fire extinguisher can be an effective tool. The basic steps for using a fire extinguisher are:

- identify the type of material involved in the fire
- select the appropriate fire extinguisher
- test the extinguisher (severe burns could result if you get to the fire and find an empty or defective extinguisher in your hand)
- approach upwind of the fire; this will keep the heat, smoke and extinguishing agent from blowing back on you

- aim at the base of the flames and use a sweeping motion to extinguish the fire
- back away from the fire before the extinguisher is emptied. Never turn your back on a fire

8.8 FIRE TYPES AND CHARACTERISTICS

Some of the safety equipment used to prevent the spread of a fire include: fire monitors, fire extinguishers, fire hose stations, sprinkler systems, and CO_2 and Halon systems.

Liquid fires depend on a gas and air mixture above the product. The liquid itself does not burn but the vapor-rich air above the liquid ignites and burns. Heat increases the vaporization process of the product by stimulating the activity of the molecules. The higher the temperature, the higher the flammable vapor concentrations will be above the liquid. Most liquid fires are pool type. Pool type configurations can be found in a tank, diked area, unconfined spill (river, lake, or land), or leaking mechanism. In contrast to a gas fire, a liquid fire's burning rate is dictated by how fast the flame can transfer heat back to the liquid.

In solid fuel fires heat is fed back to the solid fuel and a complex gasification process takes place. The individual characteristics of the solid fuel complicates heat release estimates. Solid fuels tend to thermally decompose rather than evaporate. The heat of gasification for solids includes the heating of the material, melting, and the heat of decomposition. Fire fighters use the same method for calculating pool fire burning rates on solid fuel fires.

As mentioned earlier, solid and liquid fires must burn in the gas phase. Liquid and solid fires are limited in burn rate by heat transfer to the source. Gas fires are usually jet flames. Jet flames usually result when gas escapes from its special containment. Gas release rates and heat release rates are determined by the chemical characteristics of the gas. The only difference between liquid, solid, and gas fires is the method and rate of fuel gas evolution.

Chemical explosions take place when high pressure gas is discharged from the generation of chemical reactions. There are five basic types of chemical explosions:

1. Combustion explosions—this type includes dust, gas, mists, and backdrafts. An unconfined vapor cloud explosion (UVCE) is included in this group.
2. Thermal explosions—two or more chemical compounds combine to rapidly react and explode. Common to petrochemical and refinery operations.
3. Condensed phase explosions—rapid chemical component reactions between solids or liquids. Associated with high explosives.
4. Nuclear explosions—fission and fusion of matter.
5. BLEVE—a physical explosion; a boiling liquid expanding vapor explosion, involving the release and ignition of an expanding vapor (hence, BLEVE).

When a fire starts in a leak tight room, substantial pressures are created by expanding gases. Fire growth and relative tightness of the room determine total pressures. Heating and expansions normally are linked. Usually, this maximum pressure is never achieved because excessive pressure escapes easily from normal structures. Compartment fire pressures are closely linked to buoyancy forces and not to expansion of hot gases.

Once a fire has started, heat is fed back to the fuel and the fire begins to spread. As a smoke plume forms, thermal radiation continues to heat the compartment. The warm smoke plume entrains cool air which increases upward flow rates. As the plume strikes the ceiling, it forms a ceiling jet. The ceiling jet is distributed equally across the ceiling and flows until it contacts a wall. At this point, the room begins to fill with smoke from the ceiling down. Total plume gas flow is dependent upon fire size and height.

- Zone 1: ceiling—hot uniform
- Zone 2: floor—cool uniform
- Zone 3: gas temperature uniform from 2.1 meters to 1 meter

As the smoke descends it will select the first available release or vent to the outside (such as a door or a window). Flow rate is dependent upon the smoke layer size, temperature, and vent size. Ambient pressure increases from the top of the room to the floor. The weight of the air increases with the column depth similar to the hydrostatic pressure of the water in an elevated tank. The higher the column of water, the higher the pressure at the base of the tank.

Fire growth is characterized by five (5) stages:

1. Ignition—fire starts
2. Growth—fire size increases by using up initial combustible material
   - radiation stage (vigorous burning)
   - enclosure stage (reached when gas layer at ceiling and other heated objects reradiate substantial heat back toward the fuel surface)
   - ceiling stage (flames reach ceiling; heat is reradiated)
3. Flashover—500°C to 600°C; entire room is on fire
4. Fully developed fire—steady burning of combustibles
5. Decay—fuel begins to be consumed; temperature declines consistently.

The fire plume is located directly above the fire. As the fire burns, hot gases are released into the atmosphere. As these gases rise, they entrain cooler air which speeds upward velocity. Fire plumes constantly change during the life of the fire in size, temperature, and upward flow rate. The type of fire also determines the characteristics of the plume: turbulent, hydrocarbon, wood, and so on.

Fire growth and shape are dependent upon the materials that made up the fire during the ignition stage. At the start of the growth stage, the fire begins to burn vigorously. At this point combustible materials in the room are exposed to excessive heat. Low density, thin, vertical materials will be consumed easily. If the fuel in the fire is a flammable gas or liquid then the growth rate will be enhanced. Sustained temperature increases or decreases will come from the physical properties of the combustible material.

A liquid hydrocarbon fire is characterized by high temperatures and emissivity rates. Since this type of fire has a higher heat release rate than a typical wood fire, its plume is taller and more turbulent and the flames reach much higher. Hydrocarbon fires are also more luminous because of the amount of soot they have entrained in the flames.

The shape of the plume is dependent upon plume velocity and entrainment. The flow entrained is proportional to the plume velocity at a particular elevation. Flame growth and shape are affected by:

- sudden wind changes, which increase flame exposure to combustible materials (concurrent flame spreads)
- dry conditions
- variety in chemical components consumed by the fire
- type of fire: gas, liquid, or solid

SUMMARY

The fundamentals of fire prevention, protection, and control include analyzing the physical layout of the plant or facility and surrounding structures, ensuring that fire fighting equipment is available and strategically placed, and designing engineering systems that allow for the safe handling, storage and shipment of flammable and combustible materials. Written procedures and emergency response plans, structured safety training programs, and early fire detection systems will help prevent the outbreak and spread of fires.

The development of a fire can be traced through the incipient stage, the smoldering stage, the flame stage, and the heat stage. Fire growth is characterized by five (5) stages: ignition, growth, flashover, fully developed fire, and decay.

Fire is a chemical reaction between oxygen, fuel, and heat. If any one of these three elements is removed, the fire will go out. In the fire triangle, fuel must be present in vapor form, must be in the flammable range, and there must be oxygen or an oxidizing agent; and heat must be in the form of high environmental temperatures, hot surfaces, open flames, friction, and so on.

The fire classification system is designed to simplify the selection of fire fighting techniques and equipment. Class A fires involve the burning of combustible materials such as wood, paper, plastic, cloth fibers, and rubber. Class B fires involve combustible and flammable gases and liquids and grease. Class C fires are categorized as electrical fires. This involves energized equipment and Class A, B and D materials that are located near the fire. Class D fires cover combustible metals such as aluminum, magnesium, potassium, sodium, titanium, and zirconium.

The five most common fire extinguishers found in the chemical processing industry include CO_2 extinguishers (effective on Class B and C fires), dry chemical fire extinguishers (effective on Class A, B and C fires), foam fire extinguishers (used to control flammable liquid fires and are effective on Class A and B fires), Halon fire extinguishers (designed for use on Class A, B, and C fires), and water fire extinguishers (designed for use on Class A fires only).

The basic rules for using a fire extinguisher are to identify the type of material involved in the fire, select the appropriate fire extinguisher, test the extinguisher, approach upwind of the fire, aim at the base of the flames, use a sweeping motion to extinguish the fire, and back away from the fire before the extinguisher is emptied.

There are five basic types of chemical explosions: combustion, thermal, condensed phase, nuclear, and BLEVE.

REVIEW QUESTIONS

1. Name the three components of the "fire triangle."
2. Your computer console has just caught on fire. List the steps you would take to put the fire out. Identify the type of extinguisher you would use.
3. The motor on pump 102 has just caught on fire. List the steps you would take to put it out. Identify the type of extinguisher you would use.
4. A liquid spill from tank 102 has just caught on fire. Tank 102 is empty. List the steps you would take to extinguish the fire. Identify the type of extinguisher you would use.
5. Fire monitors are used to _______ exposed facilities and equipment and limit the spread of the fire. The fire monitor is equipped with an adjustable _______ gallon per minute.
6. The nozzle on a fire monitor has three settings. Name them.
7. List the four stages of a fire.
8. How do you report a fire?
9. Describe three principles of fire prevention, protection, and control.
10. Describe the fire classification system.
11. Describe the hazards associated with fighting a fire that has the potential to produce a BLEVE.

Chapter 9

HAZWOPER

Government Mandate:

HAZWOPER 29 CFR 1910. 120

OBJECTIVES

- ***Describe the HAZWOPER first responder awareness level.***
- ***Describe the HAZWOPER first responder operations level.***
- ***Describe emergency response.***
- ***Identify potential hazards encountered during an emergency situation and the impact each could have.***
- ***Describe the incident command system.***

KEY TERMS

Autoignition—the temperature at which a liquid will spontaneously ignite without a spark or flame.

Awareness level—an individual who is trained to respond to a hazardous substance release, initiate an emergency response, evacuate the area, and notify proper authorities.

Boiling point—the temperature at which a liquid changes to a vapor.

CFR—Code of Federal Regulations.

Degradation—the breakdown or loss of physical properties.

Emergency response—a loss of containment for a chemical or the potential for loss of containment that results in an emergency situation requiring an immediate response.

HAZMAT—Hazardous materials.

HAZMAT response team—team of people who have received specialized training so that they can control and handle a hazardous chemical spill or release.

Incident commander—person responsible for organizing and coordinating response activities.

Incident command system—a military-type system designed to respond to an emergency.

Loss of containment—occurs when a chemical escapes from a vessel, pipe, or piece of equipment.

Operations level—an individual who is trained to respond with an aggressive posture during a chemical release by going to the point of the release and attempting to contain or stop it.

Oxidizer—a chemical that yields oxygen.

Pyrophoric—ignites spontaneously in air below 130°F.

Safe Haven—a designated area that is safe from vapor releases.

Vapor pressure—the pressure exerted in a confined space by the vapor above its own liquid.

9.2 **HAZWOPER** 29 CFR 1910.120

HAZWOPER is used to describe OSHA's Hazardous Waste Operations and Emergency Response standard. The standard covers two important parts of a plant's operation: emergency response and hazardous waste operations. The 29 CFR 1910.120 requires that all individuals who respond to an emergency situation have at least 24 hours of training. Refresher training is also covered under the standard while plant specific requirements may include additional training. HAZWOPER is broken down into the following areas:

- Emergency response—first responder awareness level, first responder operations level, hazardous materials technician, and specialist level
- Hazardous waste operations—incident command system, scene safety and control, spill control and containment, decontamination procedures, emergency termination, or all clear
- Hazard protection, prevention and control—terms and definitions, PPE levels, identifying hazardous materials, hazards initiating an emergency response, avoiding hazards, entry of hazardous materials into the body, unit monitors, and field survey instruments

The HAZWOPER standard defines five levels for emergency responders. These five levels include:

1. first responder awareness level
2. first responder operations level
3. hazardous materials technician level
4. hazardous materials specialist
5 on scene incident commander

NOTE: The information in this chapter will reflect objectives specific to the first responder awareness level and first responder operations level.

First Responder Awareness Level

Process technicians play a key role in the operation and maintenance of chemical plants and refineries. Due to their close proximity to the operation and the chemicals found in the process, they are very likely to be the first to witness a release. The first responder awareness level is directed at individuals who witness or discover a hazardous chemical release and who have received emergency response training. Properly trained technicians know how to recognize a hazardous chemical release, the hazards associated with this release, how to initiate the emergency response procedure, and how to notify appropriate personnel.

First Responder Operations Level

HAZWOPER 29 CFR 1910.120 training is required upon initial assignment and annually each year thereafter. When a hazardous chemical release occurs, the process technicians working

Figure 9.2-1 *HAZWOPER*

on the unit specific to the release will attempt to respond. Their efforts are directed at preventing the spread of the release and at saving lives, equipment, the community, and the environment.

NOTE: The process technicians operating a unit will know more about that specific process than anyone else. It is virtually impossible to pull in a highly trained emergency response team who would know the unit a fraction as well as the technician. As long as the situation is under minimal control, the process technician is the best first line of defense.

First responders for the operations level learn:

- how to respond to, control, contain, and confine a release
- communication procedures between plant personnel
- structure of emergency response system
- how to use personal protective equipment
- standard emergency response and termination procedures

Events that Would Trigger an Emergency Response

- loss of containment
- punctured 55 gallon drum
- pump or compressor seal failure
- overflowing tank
- pipe or vessel leak
- explosion or fire
- gas release or vapor release
- toxic chemical spill or release

9.3 EMERGENCY RESPONSE

The chemical manufacturing industry defines emergency response as a loss of containment for a chemical or the potential for loss of containment that results in an emergency situation requiring an immediate response. Examples of emergency response situations include fires, explosions, vapor releases, and reportable quantity chemical spills.

Refinery and petrochemical plant employees who are likely to discover or witness a chemical release fall under the scope of the awareness level while those employees who take preventive measures to control and secure the release fall under the guidelines of the operations level. Emergency response procedures are applied to every individual working for a company. The chemical processing industry typically has the following groups working in and around their facilities:

- process and lab technicians
- maintenance technicians (instrument, electricians, mechanics, and so on)
- construction and janitorial technicians
- engineers and chemists
- management and administrative staff
- safety and security personnel

Each of these groups are trained to respond to an emergency situation. During an emergency situation, process technicians who fall into the first responder awareness level are required to complete specific procedures. They are trained to respond to a hazardous substance release, initiate an emergency response, evacuate the area, and notify proper authorities. For example: On July 4th, 1999, a process technician was completing a routine checklist when she noticed a large vapor cloud escaping from the top of a nearby unit's loop reactor. The chemicals escaping from the reactor were composed of an extremely flammable and explosive material. The technician recognized the potential hazard and the need to initiate her plant's emergency response plan. The following steps were taken during the emergency:

- the technician immediately radioed her supervisor about the release.
- the first line supervisor notified the FLS at the loop reactor about the problem (they were already aware of the situation because of process instrumentation and were responding. They confirmed that the situation was serious.)
- the loop reactor FLS initiated the emergency response situation. The plant alarm system for a vapor release was sounded. The emergency response teams mobilized.
- process technicians from each unit began the evacuation of their units to the designated safe havens.
- roll was taken and reported to a central incident commander. All plant personnel were accounted for.
- technicians trained to respond to the release were mobilized.
- when the release was contained and dispersed, the all clear was sounded.

9.4 HAZARDOUS WASTE OPERATIONS

Hazardous waste operations in the chemical processing industry involve the use of a complex incident command system (ICS). The ICS coordinates all emergency response activities. Under the ICS, a scene safety and control system is coordinated. Decontamination, spill control and containment, and an emergency termination program also fall under the incident command system.

The ICS is led by an incident commander. The incident commander is responsible for organizing and coordinating response activities and is surrounded by a formal organization with defined lines of authority and responsibility that provide information and carry out orders.

Hazardous waste operations are classified into three categories:

1. small hazard that does not affect the whole unit
2. medium size hazard that impacts one or more operating units
3. large hazard that impacts the plant and community

The organizational structure for a typical incident command system is:

- incident commander—directs on-scene activities
- on-scene commander—establishes command post and first response activities
- scene specialist—conducts risk assessment
- planning chief—coordinates unit operations
- operations chief—directs HAZMAT and fire fighting
- medical chief—provides industrial hygiene and first aid
- employee welfare chief—provides financial and material assistance to the community

- logistics chief—refuels vehicles; delivers supplies
- security chief—secures scene and maintains order
- communications chief—initiates radio, telephone, and communications
- public affairs chief—provides information to the media
- spill containment, clean-up chief—coordinates containment and clean up
- safety chief—assesses hazard and prevents unsafe acts

Scene Safety and Control

In order to reduce the possibility of accidental spread of hazardous chemicals into areas outside of the affected unit, a three zone system has been established.

1. hot zone—the area around the incident where contamination has occurred. Emergency response activities that occur in the hot zone require the appropriate PPE and the buddy system.
2. warm zone—used to decontaminate technicians leaving the hot zone.
3. cold zone—a staging area where the incident command post is established.

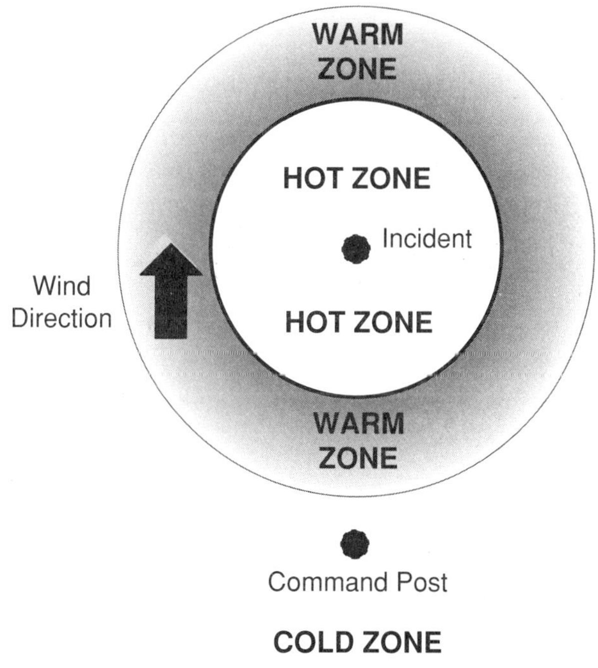

Figure 9.4-1 *Scene Safety and Control*

Spill Control and Containment

Chemical spills and vapor releases require different containment procedures. Most operating units have been designed with spill control or containment in mind. Absorbents, adsorbents, curbing and diking, segregated sewers, earthen dams, and fire monitors are devices used to control chemical spills and releases.

Decontamination Procedures

In a decontamination procedure, anything that comes into contact with a hazardous chemical is neutralized in the warm zone. The typical way of doing this is through disposal of the contaminated PPE. Contaminated workers are first rinsed off with fresh water before doffing their protective equipment. Dirty clothing is placed in a pile while dirty tools are placed in plastic bags for solid waste collection and clean up.

Hazard Identification

During a chemical spill or release, the identification of the material is important. Since time is critical during an emergency response situation, the following procedure must be completed quickly and from a safe distance.

- determine the location of the release (a number of chemicals can be identified through this procedure)
- analyze the composition of the material being released (determine whether it is a gas, vapor, liquid, solid, or a combination)
- identify other characteristics such as odor, color, and physical characteristics
- report and discuss findings with area technicians and supervision

Hazards Initiating an Emergency Response

Hazards that will initiate an emergency response have been determined by the chemical processing industry to be an explosion, fire, vapor release, toxic chemical release, large product or chemical spill, and loss of containment for radioactive material.

Loss of Containment

The chemical processing industry has defined the following situations as the primary causes for loss of containment: pipe or flange failure, pump seal failure, explosions, fires, overfilled tanks, and overpressured tanks, and overturned drums or containers.

Unit Monitors and Field Survey Instruments

Draeger pumps are portable vacuum-type devices used to collect a representative gas or vapor sample to determine if a known contaminate is present. The Draeger pump has a bellows to draw air into a glass tube filled with a chemically treated sorbent. The activated media in the glass tube will change color if the suspected chemical is present in the air. Each glass tube is scaled or graduated so that it can be read in parts per million. Process technicians need specialized training in order to operate a Draeger vacuum pump properly.

Explosimeters, commonly called combustible gas meters, are used to determine whether there is sufficient concentrations of a combustible gas mixture to produce a fire or explosion. Explosimeters operate by drawing in an air sample into a combustion chamber and igniting it. If the air sample ignites, a detectable combustion level registers.

Fixed unit monitors are designed to test for combustible or toxic substances such as hydrogen sulfide or chlorine. Organic vapor analyzers and oxygen analyzers are used to detect hydrocarbons and oxygen levels. Fixed detectors are strategically located around the unit and will send an audible alarm to the control room in the event that a target substance is detected.

Geiger Counters are used to measure the level of radioactivity in an atmosphere by counting the number of ions given off by a source and the rate at which they are lost.

Bunker Gear

Several years ago, a fire broke out in a plant which required the process technicians to call in a fire fighting team to assist. After a short time, an explosion occurred which resulted in a huge fire ball that engulfed several technicians and fire fighters. The heat from the fire immediately killed the technicians but inflicted only minor injuries on the fire fighters. Upon closer investigation, it was determined that the bunker gear that the fire fighters were wearing provided superior protection over the flame retardant clothing (FRC) that the technicians were wearing.

Requirements of the Bunker Gear standard fall under 29 CFR 1910.120 and are provided for the purpose of training individuals who will respond to a major fire or vapor release with a fire team. Most chemical companies and refineries are giving new employees this training upon initial assignment and annually thereafter. Bunker gear PPE consists of:

- gloves—special heat resistant gloves
- boots—special heat resistant material
- coats—special heat resistant material; come in small, medium, large and extra large; all latches are engaged, none can be left open; the collar should be pulled up straight and the strap fastened
- helmet—special heat resistant material with a face shield; inner liner should be pulled down over the ears and the chin strap secured. A rotary adjustment knob is located on the back of the helmet to ensure that the helmet fits snugly on the head.

SUMMARY

The term HAZWOPER is used to describe OSHA's Hazardous Waste Operations and Emergency Response standard. HAZWOPER is broken down into the following areas: emergency response, hazardous waste operations, and hazard protection, prevention, and control.

Emergency response is the loss of containment for a chemical or the potential for loss of containment that results in an emergency situation requiring an immediate response. The levels of response have been determined by OSHA to be:

1. first responder awareness level
2. first responder operations level
3. hazardous materials technician level
4. hazardous materials specialist
5. on-scene incident commander

Hazardous waste operations in the chemical processing industry involve the use of a complex incident command system (ICS). The ICS coordinates all emergency response activities. In order to reduce the possibility of accidental spread of hazardous chemicals into areas outside of the affected unit, a three zone system has been established: hot zone, warm zone, and cold zone.

REVIEW QUESTIONS

1. What is the purpose of HAZWOPER?
2. Describe why a process technician needs to have emergency response training.
3. The chemical processing industry has five basic levels of emergency response. Name them.
4. Contrast the first responder awareness level with the operations level.
5. How does the ICS work for minor, medium and major vapor releases and fires?
6. Name the hazards that will initiate an emergency response.
7. Define emergency response.
8. Describe hazardous waste operations.
9. In order to reduce the possibility of the accidental spread of hazardous chemicals into areas outside of the affected unit, a three zone system has been established. List the zones.

Chapter 10

Environmental Standards

OBJECTIVES

- *Describe key elements of environmental awareness training.*
- *Describe air pollution control.*
- *Describe water pollution control.*
- *Describe solid waste control.*
- *Describe toxic substances control.*
- *Describe the Resource Conservation and Recovery Act (RCRA).*
- *Explain the purpose of the Toxic Substances Control Act.*
- *Contrast the clean up of chemical waste and the Comprehensive Environmental Response, Compensation and Liability Act CERCLA or Superfund Law.*
- *Describe the roles and responsibilities of the EPA.*

KEY TERMS

Air permits—permits obtained for any project that has the possibility of producing air pollutants.

Air pollution—the contamination of the air especially by industrial waste gases, fuel exhausts, or smoke.

Clean Air Act—the purpose of the Clean Air Act is to (1) enhance the quality of the air; (2) accelerate a national program to prevent air pollution; (3) provide technical and financial assistance to state and local government; (4) develop a regional air pollution control program.

Clean Water Act—regulates the discharge of pollutants into lakes, streams, and oceans.

Community right-to-know—increases community awareness of the chemicals manufactured or used by local chemical plants and business; involves the community in emergency response plans, improves communication and understanding, improves local emergency response planning, and identifies potential hazards.

Emergency response—a procedure initiated by the loss of containment for a chemical or the potential for loss of containment that results in an emergency situation requiring an immediate response. Emergency response drills are carefully planned and include preparations for worst case scenarios. Examples include vapor releases, chemical spills, explosions, fires, equipment failure, hurricane, high winds, loss of power, and bomb threats.

Resource Conservation and Recovery Act (RCRA)—enacted as Public Law in 1976; the purpose of RCRA is to protect human health and the environment. A secondary goal is to conserve our natural resources.

Solid waste—a by-product of modern technology. Solid waste is technically defined as a discarded solid, liquid, or containerized gas. This definition includes materials that have been recycled or abandoned through disposal, burning or incineration, accumulation, storage, or treatment.

Toxic Substances Control Act (TSCA)—a Federal Law enacted in 1976 intended to protect human health and the environment. TSCA was also designed to regulate commerce by requiring testing and placing necessary restrictions on certain chemical substances. TSCA imposes requirements on all manufacturers, exporters, importers, processors, distributors, and disposers of chemical substances in the United States.

Water pollution—the contamination of the water especially by industrial wastes.

10.2 WATER POLLUTION (Exxon Valdez)

On March 24, 1989, the Exxon Valdez ran aground on a reef in Prince William Sound, Alaska. Approximately 11 million gallons of North Slope crude oil gushed out of the ruptured hull contaminating the coastlines of the Alaska Peninsula, lower Cook inlet, the Kenai Peninsula, the Kodiak Archipelago, and Prince William Sound. Oil from the Valdez was found over 600 miles southwest of the spill. The spill had catastrophic effects on five State Parks, three National Parks, one State Game Sanctuary, a National Forest, four National Wildlife Refuges, and four State Critical Habitat Areas.

The Valdez incident was the largest tanker oil release in U.S. history. Cleanup and containment efforts started immediately but five years after the spill, oil could still be found in sensitive areas. During the cleanup efforts, thousands of people responded. A fleet of private fishing vessels went to the aid of critical fishing hatcheries by setting up an outer perimeter of skimmers to remove the oil. Exxon initiated a massive cleanup effort and funded millions of dollars for recovery efforts. Workers cleaned beach fronts with high pressure steam and bioremediation techniques.

Figure 10.2-1 *Exxon Valdez*

10.3 AIR POLLUTION

Modern factories, office buildings, and homes produce smoke that contains sulfur dioxide and nitrogen dioxide. Sulfur dioxide and nitrogen dioxide are known to have harmful effects on the human respiratory system. As sulfur dioxide travels higher into our atmosphere, it turns into sulfur trioxide that mixes with rain to form sulfuric acid. As nitrogen dioxide rises it combines with hydrocarbons and sunlight to form smog. Smog is an irritant that damages the eyes, the respiratory tract, and plants. Motorized vehicles produce unburned hydrocarbons, carbon monoxide and nitrogen dioxide. This is bad for our environment because hydrocarbons injure plant life and human life and help form smog. Carbon monoxide causes headaches and dizziness in humans. High concentrations of carbon monoxide can kill. Incineration units and wide scale burning concentrates high levels of mercury and small particles in the atmosphere. This reduces visibility, injures the respiratory tract and nervous system, and affects the climate.

Modern technology produces a variety of useful products. This same technology produces bi-products that can harm the environment. Because of the potential hazards that exist with new technology, environmental laws and regulations have been passed to protect our future, such as the Clean Air Act. The purpose of the Clean Air Act is to enhance the quality of the air, accelerate a national research and development program to prevent air pollution, provide technical and financial assistance to state and local government, and develop a regional air pollution control program.

Air Pollution Control

In 1955, the original clean air act was passed. Since 1955, a number of modifications have been made.

- 1960 amendment directed the Surgeon General to study vehicle pollution
- 1963 amendment directed research into fuel desulfurization and development of air quality criteria
- 1965 amendment studied new sources of pollution
- 1967 Quality Air Act passed
- 1970 Clean Air Amendment
- 1977 amendment to the Clean Air Act for emission standards
- 1990 reauthorization of Federal Clean Air Act, including:
    - air toxins
    - acid deposition
    - job training for laid-off workers due to Clean Air Act
    - air quality standards
    - permits
    - stratospheric ozone and global climate protection
    - provisions for enforcement
    - acid rain and air monitoring research
    - provision to improve air quality and visibility near national parks
    - provisions relating to mobile sources

Agencies

The EPA was established in 1970 to protect the environment from pollution. The creation of the EPA brought fifteen federal programs under one umbrella and is most noted for its management of the Comprehensive Environmental Response, Compensation and Liability Act of 1980 (Superfund). In 1983, the EPA received severe criticism from Congress for their handling of Superfund. The synergy created by Congress and special interest groups made it clear that the public opinion about a clean environment vetoed the interests of the chemical processing industry. The enforcement of these new, tough, environmental regulations put many industrial manufacturers out of business.

Initially, the EPA focused on recycling and cleaning up open dump sites. Today, the government has passed over twelve environmental laws that impact air, water, and land. These laws go from cradle to grave or, in other words, they hold chemical manufacturers responsible from the time the chemical is created until it is disposed of. The EPA regulates water quality, pesticides used on farms, oil spills, everything that goes into the ground, and many other things.

Figure 10.3-1 *Environmental Awareness*

The Air Pollution Control Board maintains numerous regional offices throughout the United States. Each location receives public complaints, coordinates investigations, documents violations, and recommends enforcement actions.

Air Permits
Permits must be obtained for any project that has the possibility of producing air pollutants. The Air Control Board (ACB) places limits on emissions and needs about three to eight months to complete the permit process. After the ACB issues the permit, a yearly inspection is scheduled. Penalties for civil and criminal abuses of the Clean Air Act range from $25,000 a day to $250,000 and two to fifteen years in jail. For example, smoking flares for longer than five minutes is illegal and should be reported. Failure to report this crime results in severe penalties.

10.4 WATER POLLUTION CONTROL

The Federal Clean Water Act was passed in 1898. Fifty years later, Congress provided funds for the construction of municipal waste water treatment facilities. The Water Control Act of 1965 took a "water quality" approach and initiated close examination of receiving waters. States were required to establish standards for water quality. The Clean Water Act of 1972 adopted the Best Available Technology (BAT) strategy for all cleanups. Under the 1987 amendments, states are required to identify waters that are not expected to meet quality standards.

Water Pollution Controls
The Clean Water Act regulates waste water. Waste water standards are applied to:

- process waste water, process contact water, contaminated water from vessels and equipment, tanks, slab cleanup, and so on
- rainwater and sewer system releases
- once-through cooling water, cooling tower blow-down, and boiler blow-down

The Federal Clean Water Act is designed to protect United States water quality. The EPA, State Water Commission, Army Corps of Engineers, State Parks and Wildlife, and U.S. Fisheries and Wildlife help enforce the Clean Water Act.

National Water Quality Standards

- all United States waters shall be fishable and swimable
- no discharge of toxic pollutants in toxic quantities will be allowed
- technology must be developed to eliminate pollutant discharge

Water Permits
The Clean Water Act requires a company to have a water permit. In some states, a two permit system exists.

10.5 SOLID WASTE CONTROL

Solid waste is a by-product of modern technology. Solid waste is technically defined as a discarded solid, liquid, or containerized gas. This definition includes materials that have been recycled or abandoned through disposal, burning or incineration, accumulation, storage, or treatment.

The Resource Conservation and Recovery Act (RCRA) was enacted as Public Law in 1976. The purpose of RCRA is to protect human health and the environment. RCRA focuses on the correct disposal of hazardous wastes, recycling practices, and solid waste disposal. A secondary goal of RCRA is to conserve our natural resources. RCRA completes this task by regulating all aspects of hazardous waste management including generation, storage, treatment, and disposal. This concept is referred to as "cradle to grave."

The RCRA includes a civil penalty of $25,000 a day, a criminal penalty for willful endangerment of $250,000 plus fifteen years in jail, and a $1,000,000 fine for a company. Liability includes any person involved in breaking the law.

Solid waste categories:

- Class One: Hazardous—ignitable, reactive, corrosive, toxic
- Class One: Non-hazardous—RCRA regulations do not apply
- Class Two: Garbage, cured epoxy resin, biopond filter solid
- Class Three: uncontaminated or inert material such as wood

The cleanup of waste sites is regulated by the Comprehensive Environmental Response Compensation and Liability Act (CERCLA). This act is frequently referred to as the Superfund law. Under CERCLA, a number of large companies have been required to provide money for cleanup activities for materials dumped by them or their parent companies years ago. This effort supports the cradle to grave concept and forces chemical manufacturers to develop long range procedures for the handling, storage, and disposal of chemicals. The State Water Commission is the primary regulating agency charged with enforcement of solid waste disposal generated within their boundaries.

Permits

More than ninety day storage of a hazardous chemical requires a permit. An ideal facility would include:

- a covered facility to prevent rain water contamination
- no contact with soil
- containment for all equipment
- raised equipment to permit inspection for leaks

10.6 TOXIC SUBSTANCES CONTROL

The Toxic Substances Control Act (TSCA) is a federal law enacted in 1976. TSCA was intended to protect human health and the environment and was designed to regulate commerce by requiring testing and necessary restrictions on certain chemical substances. TSCA imposes requirements on all manufacturers, exporters, importers, processors, distributors, and disposers of chemical substances in the United States.

Controls

The TSCA inventory has over 70,000 chemicals listed and was established to record all products manufactured, imported, sold, processed, or used for commercial purposes. Exemptions include research and development chemicals and by-products without commercial purpose. TSCA also controls pre-manufacture review of new chemical substances, risk assessment by testing and information gathering, record keeping and reporting on health and environmental effects associated with chemical substances, and restrictions on known hazardous chemicals.

The TSCA has severe penalties for those who break the law. Yearly penalties for current violations are estimated at over $40,000,000. The EPA is the primary agency charged with enforcing toxic substance control.

10.7 COMMUNITY RIGHT-TO-KNOW

In 1986, Congress enacted the Emergency Planning and Community Right-To-Know Act (EPCRA). EPCRA is designed to open the lines of communication between fire departments, hospitals, refineries and chemical plants, local government, and interested community members. The idea is to establish a system that will provide the community with information about chemicals used near them and a national system for reporting accidental releases. EPCRA also provides a structured system for responding to emergency situations. Under this system, emergency responders would have access to MSDS and other specific information so they would know how to respond to each emergency.

SUMMARY

Air pollution is defined as the contamination of the air especially by industrial waste gases, fuel exhausts, or smoke. Water pollution is defined as the contamination of the water especially by industrial wastes. Solid waste is technically defined as a discarded solid, liquid, or containerized gas. Solid waste is a by-product of modern technology. This definition includes materials that have been recycled or abandoned through disposal, burning or incineration, accumulation, storage, or treatment. In an industrial environment, responding to an emergency follows a specific set of standards. Emergency response drills are carefully planned and include preparations for worst case scenarios. The community right-to-know principle:

- increases community awareness of the chemicals manufactured or used by local chemical plants and business
- involves community in emergency response plans
- improves communication and understanding
- improves local emergency response planning
- identifies potential hazards

The RCRA was enacted as Public Law in 1976. The purpose of RCRA is to protect human health and the environment. A secondary goal is to conserve our natural resources. RCRA completes this goal by regulating all aspects of hazardous waste management including generation, storage, treatment, and disposal. This concept is referred to as cradle to grave. The TSCA is a federal law enacted in 1976 to protect human health and the environment. TSCA was also designed to regulate commerce by requiring testing and necessary restrictions on certain chemical substances. TSCA imposes requirements on all manufacturers, exporters, importers, processors, distributors, and disposers of chemical substances in the United States.

REVIEW QUESTIONS

1. Define solid waste.

2. The community right-to-know principle is designed to:

3. The purpose of RCRA is:

4. Vapor releases, chemical spills, explosions, fires, equipment failure, hurricane, high winds, loss of power, and bomb threats fall under which main program?

5. Cradle to grave is a term associated with which government act?

6. Smoking flares in excess of ___ minutes should be reported. Failure to report results in severe penalties.

7. ______________________ must be obtained for any project which has the possibility of producing air pollutants.

8. What was the significance of the Exxon Valdez incident?

9. List the harmful products produced by air pollution. Include how they are produced.

Chapter 11

Selected Topics

Government Mandates:

Occupational Noise Exposure 29 CFR 1910.95
Asbestos Awareness and Gasket Removal 29 CFR 1910.1200/1001 and 29 CFR 1926.59/1101
Bloodborne Pathogens 29 CFR 1910.1030

OBJECTIVES

- ***Describe the hearing conservation standard.***
- ***Explain the effects of noise on hearing.***
- ***Review the purpose of hearing protection devices.***
- ***Describe the selection, use, fit, and care of hearing protection.***
- ***Explain the purpose of audiometric testing.***
- ***Describe the bloodborne pathogens standard.***
- ***Describe the background and history of asbestos.***
- ***Describe a fall protection system.***
- ***Describe the equipment used in fall protection.***
- ***Identify safe work practices used with fall protection.***
- ***Complete a fall protection inspection checklist.***
- ***Describe the proper cleaning and storage procedures associated with fall protection equipment.***

KEY TERMS

Administrative noise abatement—reduce employee exposure time to industrial noise.

Anacusis—total hearing loss.

Anchor point—a tie off connection device used to secure the free end of a full body harness lanyard.

Asbestos—a fire retardant material used in brake linings, hair dryer components, insulation, and shingles. It was used by the ancient Greeks for lamp wicks and by the Egyptians as mummy wrap.

Asbestos-containing material (ACM)—any material that contains fibrous asbestos.

Audiometry—the science of testing, measuring, and recording hearing ability.

Buckles, d rings, and snap hooks—auxiliary equipment found on the full body harness, lanyard, and anchor point.

Engineering noise abatement—reducing noise through new equipment design and innovation.

Hearing conservation program (HCP)—29 CFR 1910.95 requires employees to reduce noise in the workplace through the use of engineering and administrative controls, provide hearing protection for employees, and implement a hearing conservation program to monitor sound levels, conduct audiometer tests, and provide training to employees.

Noise—valueless, unwanted sound.

Polyester full body harness—a safety device designed to evenly distribute the forces of an accidental fall.

Polyester lanyard—a tie off rope that is attached to a full body harness.

Reducing noise at its source—a technique that utilizes engineering controls and limits employee access to high noise areas.

11.2 HEARING CONSERVATION AND INDUSTRIAL NOISE

When OSHA was enacted in 1970, federal regulations for controlling noise in the workplace were implemented. The Occupational Safety Health Administration adopted the regulations contained in the Walsh-Healy Act and applied them to all employees. These new standards were brief and had far reaching impact that could be broken down into two major components:

1. Maximum noise exposure
2. Actions that employers must take if the limits are exceeded
   - reduce noise using engineering and administrative controls
   - provide hearing protection for employees
   - implement a hearing conservation program (HCP)
     - monitor sound levels
     - conduct audiometric tests
     - provide hearing protection
     - provide training

OSHA regulation 29 CFR 1910.95 requires all process technicians who are exposed to over eighty-five decibels over an eight hour time weighted average (TWA) period to be placed in a

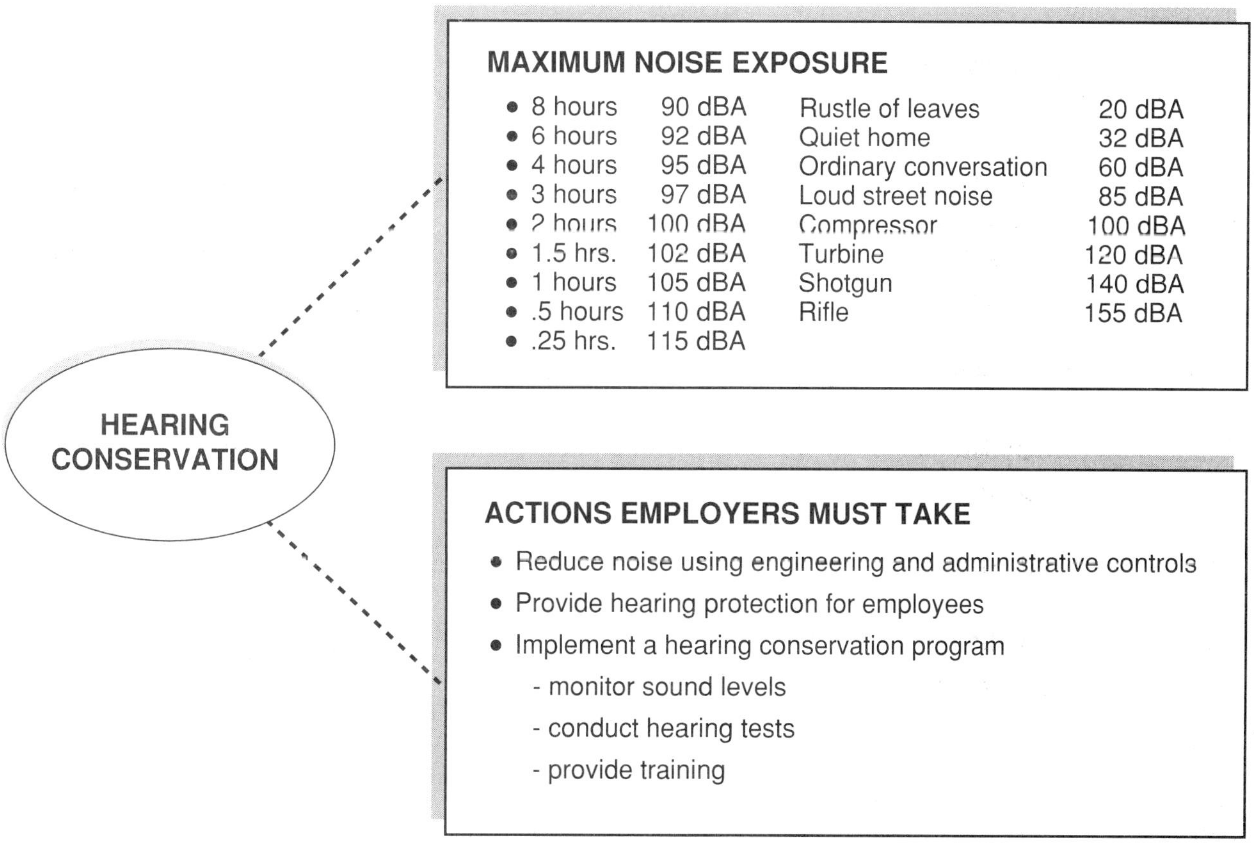

Figure 11.2-1 *Hearing Conservation*

hearing conservation program. The standard requires that each process technician receive initial training and refresher training on the:

- effects of noise on hearing
- purpose of hearing protection devices
- selection, use, fit, and care of hearing protection
- purpose of audiometric testing

Noise is loosely defined as valueless or unwanted sound. Ambient noise is the composite of all encompassing noise found in an environment. The negative effects of noise include hearing loss, fatigue, speech problems, and structural damage to the ear drum. Hearing loss occurs slowly and painlessly. Once a process technician has experienced hearing loss due to industrial noise, it is unlikely that the technician will ever recover. Approximately ten million US and Canadian workers suffer from noise-induced hearing loss.

When a compressor or pump is started, it initiates a disturbance in the air called sound. Sound is generated in two ways:

1. Air flows over a sharp edge and becomes turbulent and creates sound waves.
2. A vibrating surface sends pressure waves through a fluid or solid medium. Process equipment typically falls into this group.

SOUND WAVES

THREE DIMENSIONAL RING (SPHERE)

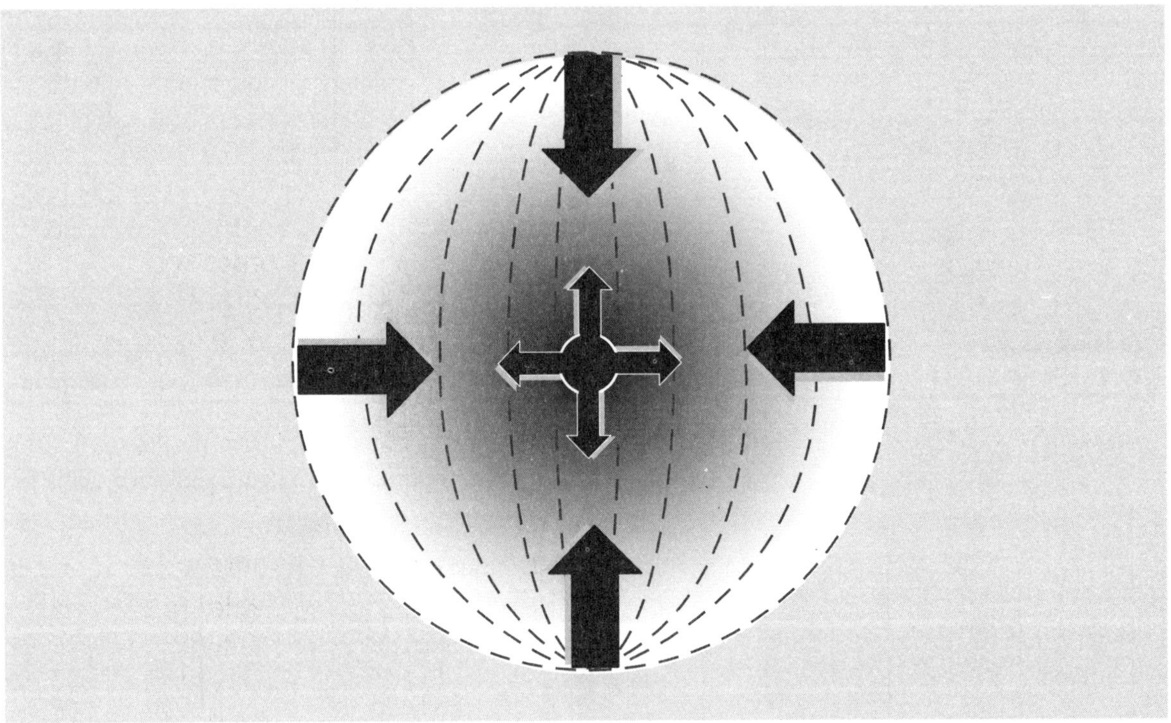

SOUND ENERGY

Transferred from one atom to the next through vibration

Figure 11.2-2 *Sound Waves Sphere*

Sound waves travel from the source of the noise outward, downward, and upward. These pressure waves roughly resemble the circular ripples a stone creates when thrown into a still pool of water. The waves easily can be seen traveling outward in a ring. The medium that the sound travels through (air, fluid, solid) does not move. Sound energy is transferred from one atom to the next through vibration. Sound waves form a three-dimensional ring that closely represents a sphere.

OSHA TABLE D-2

Permissible Exposure Limits	
Hours	**dB(A)**
8	90
6	92
4	95
3	97
2	100
1.5	102
1	105
0.5	110
0.25	115

Noise Comparison	
Rustle of Leaves	20
Quiet Home	32
Ordinary Conversation	60
Loud Street Noise	85
Compressor	100
Turbine	120
Shotgun	140
Rifle	155

Noises over 140 decibels can cause permanent hearing loss in a single exposure. Hearing loss occurs through four major categories of noise exposure:

1. overall noise level
2. intensity of noise
3. duration (in a workday)
4. duration (over a lifetime)

Audiometric testing is performed on each employee annually. When a process technician is first hired, a baseline audiometric test is run. Hearing in each ear is tested and recorded. Annual hearing tests are performed and compared to the baseline results for the rest of the technician's career.

Hearing protection can be classified in three ways:

1. Engineered noise abatement—reducing noise through new equipment design and innovation.
2. Administrative noise abatement—reducing employee exposure time to industrial noise.
3. Personal hearing protection—ear plugs, ear muffs, and expandable ear plugs are used commonly by process technicians. The earplug must be rolled between the thumb and forefinger and compressed before it can be inserted into the ear canal. The foam ear plugs will expand and form a seal between the noise and the ear drum. Ear muffs provide the same protection as ear plugs. The sealing area must be kept clear of hair, ear rings, and glasses in order to provide protection. Ear muffs and ear plugs (to be used for

noise levels of 110 decibels or more) can be used together to provide maximum protection from noise. Hearing protection is typically selected by availability, intensity of noise, and personal comfort.

11.3 BLOODBORNE PATHOGENS

The bloodborne pathogens standard became effective in March 1992. The standard is designed to protect employees who come into contact with blood or infectious materials that are suspected of containing HIV and HBV. OSHA implemented the standard to eliminate or reduce exposure through engineering controls, administrative guidelines, PPE, medical vaccinations and surveillance, employee training, and signs and labels.

The Human Immunodeficiency Virus (HIV) causes Acquired Immunodeficiency Syndrome (AIDS). The virus can only survive in the blood or body fluids environment. Transmission of HIV is through physical contact with this medium. Process technicians can come into contact with this infected medium through sexual contact, puncture wounds, blood or body fluid splash onto open injuries or the lips, mouth, nose or eyes.

The Hepatitis B Virus causes inflammation of the liver and is a major contributor of acute and chronic Hepatitis. Over 300,000 people are infected with HBV. Transmission of HBV and its flu-like symptoms is the same as HIV.

Process technicians should use recommended PPE when they are in situations where exposure could occur. Examples of PPE are safety glasses, chemically resistant gloves, pocket masks, and lab coats. Decontamination of soiled PPE is typically achieved through careful, cautious disposal.

11.4 ASBESTOS

Asbestos has been used by the ancient Greeks and Egyptians. The Greeks used asbestos for wicks in their lamps while the Egyptians used it to wrap the bodies of their dead. At the turn of the century, large scale mining and production of asbestos vaulted to commercial levels. Modern yearly consumption rates have reached 800,000 to 900,000 tons. The mineral was used in brake linings, roof shingles, insulation, fire retardant materials, and a thousand other products.

Asbestos fibers are so small that they are invisible to the human eye. When compared to the human hair, an asbestos fiber is so small that it would be like comparing a pencil to a giant redwood tree. A single asbestos fiber dropped from a height of eight feet would take forty to seventy hours to reach the ground.

When an abrasive, irritating, asbestos fiber enters the human lung it activates the natural body defense system by encapsulating it. This process creates scar tissue that will remain inside the lung for life. Over time, the lungs loose their ability to move oxygen in and out and to supply oxygen to the blood stream. This condition is referred to as asbestosisis. Since asbestos is classified as a cancer-causing agent, asbestosis is considered serious.

11.5 FALL PROTECTION

Fall protection is a standard designed to reduce or eliminate injuries from accidental falls. The standard requires fall protection to be used when a technician needs access to or works at an elevation of six feet. The only exception to this is if proper facilities for working at six feet or higher already exist. Examples of situations that would require a process technician to comply with the standard are working in suspended baskets, working in motorized lifts, standing on portable ladders, and working on a pipe rack.

The essential elements of a fall protection system include:

1. written fall procedure program
2. basic fall protection equipment
   - full body harness—a safety device designed to evenly distribute the forces of an accidental fall to the strongest muscles in the body. This protects vital organs and limits the possibility of a serious injury.
   - lanyard—a tie off rope that is attached to a full body harness. It is designed to support a minimum of 5000 pounds. The fall protection standard requires that the lanyard be a maximum of six feet long, have double locking snap hooks on each end, and include a shock absorbing device that is attached to the harness "D" ring.
   - anchor point—a tie off connection device used to secure the free end of a full body harness lanyard. Examples of anchor points include pre-engineered eye bolts, cable, slide rail, structural steel, or pipe. The anchor point should be strong enough to support 5000 pounds.
3. employee training
   - all employees should be given training on the fall protection program, donning and doffing fall protection equipment, and inspecting the equipment.
4. inspection and audit program
   - visually inspect equipment for damage. Knotted lanyards cannot be used because their tensile strength is reduced by 50%. Harnesses should be inspected for wear at the buckles and "D" ring. If fiber wear is detected or a cut or tear is found in the harness material it should be replaced immediately. Check "D" rings and buckles for sharp edges, cracks, dents, burrs, or corrosion. Check snap hooks and ensure that they operate correctly.

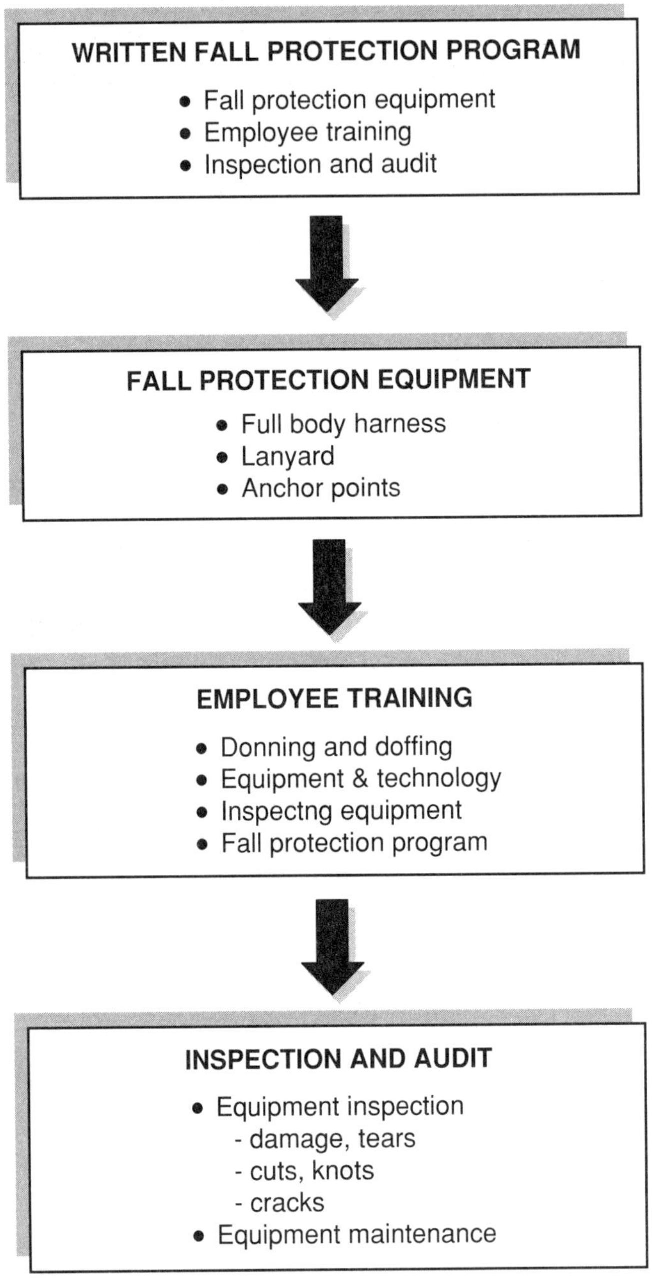

Figure 11.5-1 *Fall Protection*

After a full body harness and lanyard have been used they should be returned to their appropriate location. Harnesses can be cleaned with a wet soapy sponge.

Steps for donning a full body harness include;

1. remove harness from container and visually inspect it

2. hold harness by "D" ring and shake so straps will fall into place
3. put the harness on as you would a jacket
4. secure leg straps
5. adjust straps so the weight is transferred to the designed locations
6. visually inspect and secure the lanyard to the "D" ring double locking hook and shock absorbing device
7. secure the lanyard to the anchor point with the double locking hook as soon as you reach your work location

SUMMARY

Federal regulations for controlling noise include maximum noise exposure and actions that employers must take if the limits are exceeded. For example, reduce noise using engineering and administrative controls, provide hearing protection for employees, and implement a hearing conservation program (HCP). The HCP should monitor sound levels, conduct audiometric tests, provide hearing protection, and provide training.

Noises over 140 decibels can cause permanent hearing loss in a single exposure. Hearing loss occurs through four major categories of noise exposure:

1. overall noise level
2. intensity of noise
3. duration (in a workday)
4. duration (over a lifetime)

The bloodborne pathogens standard is designed to protect employees who come into contact with blood or infectious materials that are suspected of containing HIV and HBV. When an abrasive, irritating, asbestos fiber enters the human lung, it activates the natural body defense system by encapsulating it. This process creates scar tissue that will remain inside the lung for life. Over time the lungs loose their ability to move oxygen in and out of the lungs and supply oxygen to the blood stream. This condition is referred to as asbestosis. Since asbestos is classified as a cancer-causing agent, asbestosis is considered serious.

The fall protection standard requires fall protection to be used when a technician needs access to or works at an elevation of six feet. Fall protection is a standard designed to reduce or eliminate injuries from accidental falls. The essential elements of a fall protection system include a written fall procedure program, basic fall protection equipment such as a full body harness, a lanyard and anchor point, employee training, and an inspection and audit program.

REVIEW QUESTIONS

1. Describe the hearing conservation standard.
2. Explain the effects of noise on hearing.
3. Review the purpose of hearing protection devices.
4. Describe the selection, use, fit, and care of hearing protection.
5. Explain the purpose of audiometric testing.
6. Describe the bloodborne pathogens standard.
7. Describe the background and history of asbestos.
8. Describe a fall protection system.
9. Describe the equipment used in fall protection.
10. Identify safe work practices used with fall protection.
11. Complete a fall protection inspection checklist.
12. Describe the proper cleaning and storage procedures associated with fall protection equipment.

OSHA Regulations (Standards - 29 CFR)

Several OSHA Regulations Standards are included in this textbook as Appendices A through G, because of their importance to Process Technician's. OSHA Standards are updated and amended frequently. Readers of this textbook are cautioned to familiarize themselves with all applicable updates and amendments.

Appendix A

OSHA Regulations (Standards - 29 CFR) Process Safety Management. - 1910.119

Standard Number: 1910.119
Standard Title: Process Safety Management of Highly Hazardous Materials
SubPart Number: H
SubPart Title: Hazardous Materials

Purpose. This section contains requirements for preventing or minimizing the consequences of catastrophic releases of toxic, reactive, flammable, or explosive chemicals. These releases may result in toxic, fire or explosion hazards.
(a)
Application.
(a)(1)
This section applies to the following:
(a)(1)(i)
A process which involves a chemical at or above the specified threshold quantities listed in Appendix A to this section;
(a)(1)(ii)
A process which involves a flammable liquid or gas (as defined in 1910.1200(c) of this part) on site in one location, in a quantity of 10,000 pounds (4535.9 kg) or more except for:
(a)(1)(ii)(A)
Hydrocarbon fuels used solely for workplace consumption as a fuel (e.g., propane used for comfort heating, gasoline for vehicle refueling), if such fuels are not a part of a process containing another highly hazardous chemical covered by this standard;
(a)(1)(II)(B)
Flammable liquids stored in atmospheric tanks or transferred which are kept below their normal boiling point without benefit of chilling or refrigeration.
(a)(2)
This section does not apply to:
(a)(2)(i)
Retail facilities;

..1910.119(a)(2)(ii)
(a)(2)(ii)
Oil or gas well drilling or servicing operations; or,
(a)(2)(iii)
Normally unoccupied remote facilities.
(b)
Definitions.
"Atmospheric tank" means a storage tank which has been designed to operate at pressures from atmospheric through 0.5 p.s.i.g. (pounds per square inch gauge, 3.45 Kpa).
"Boiling point" means the boiling point of a liquid at a pressure of 14.7 pounds per square inch absolute (p.s.i.a.) (760 mm.). For the purposes of this section, where an accurate boiling point is unavailable for the material in question, or for mixtures which do not have a constant boiling point, the 10 percent point of a distillation performed in accordance with the Standard Method of Test for Distillation of Petroleum Products, ASTM D-86-62, which is incorporated by reference as specified in Sec. 1910.6, may be used as the boiling point of the liquid.
"Catastrophic release" means a major uncontrolled emission, fire, or explosion, involving one or more highly hazardous chemicals, that presents serious danger to employees in the workplace.
"Facility" means the buildings, containers or equipment which contain a process.
"Highly hazardous chemical" means a substance possessing toxic, reactive, flammable, or explosive properties and specified by paragraph (a)(1) of this section.
"Hot work" means work involving electric or gas welding, cutting, brazing, or similar flame or spark-producing operations.
"Normally unoccupied remote facility" means a facility which is operated, maintained or serviced by employees who visit the facility only periodically to check its operation and to perform necessary operating or maintenance tasks. No employees are permanently stationed at the facility. Facilities meeting this definition are not contiguous with, and must be geographically remote from all other buildings, processes or persons.
"Process" means any activity involving a highly hazardous chemical including any use, storage, manufacturing, handling, or the on-site movement of such chemicals, or combination of these activities. For purposes of this definition, any group of vessels which are interconnected and separate vessels which are located such that a highly hazardous chemical could be involved in a potential release shall be considered a single process.
"Replacement in kind" means a replacement which satisfies the design specification.
"Trade secret" means any confidential formula, pattern, process, device, information or compilation of information that is used in an employer's business, and that gives the employer an opportunity to obtain an advantage over competitors who do not know or use it. Appendix D contained in 1910.1200 sets out the criteria to be used in evaluating trade secrets.
(c)
Employee participation.
(c)(1)
Employers shall develop a written plan of action regarding the implementation of the employee participation required by this paragraph.
(c)(2)
Employers shall consult with employees and their representatives on the conduct and development of process hazards analyses and on the development of the other elements of process safety management in this standard.
(c)(3)
Employers shall provide to employees and their representatives access to process hazard analyses and to all other information required to be developed under this standard.

..1910.119(d)
(d)
Process safety information. In accordance with the schedule set forth in paragraph (e)(1) of this section, the employer shall complete a compilation of written process safety information before conducting any process hazard analysis required by the standard. The compilation of written process safety information is to enable the employer and the employees

involved in operating the process to identify and understand the hazards posed by those processes involving highly hazardous chemicals. This process safety information shall include information pertaining to the hazards of the highly hazardous chemicals used or produced by the process, information pertaining to the technology of the process, and information pertaining to the equipment in the process.
(d)(1)
Information pertaining to the hazards of the highly hazardous chemicals in the process. This information shall consist of at least the following:
(d)(1)(i)
Toxicity information;
(d)(1)(ii)
Permissible exposure limits;
(d)(1)(iii)
Physical data;
(d)(1)(iv)
Reactivity data:
(d)(1)(v)
Corrosivity data;
(d)(1)(vi)
Thermal and chemical stability data; and
(d)(1)(vii)
Hazardous effects of inadvertent mixing of different materials that could foreseeably occur.

Note: Material Safety Data Sheets meeting the requirements of 29 CFR 1910.1200(g) may be used to comply with this requirement to the extent they contain the information required by this subparagraph.

(d)(2)
Information pertaining to the technology of the process.
(d)(2)(i)
Information concerning the technology of the process shall include at least the following:

..1910.119(d)(2)(i)(A)
(d)(2)(i)(A)
A block flow diagram or simplified process flow diagram (see Appendix B to this section);
(d)(2)(i)(B)
Process chemistry;
(d)(2)(i)(C)
Maximum intended inventory;
(d)(2)(i)(D)
Safe upper and lower limits for such items as temperatures, pressures, flows or compositions; and,
(d)(2)(i)(E)
An evaluation of the consequences of deviations, including those affecting the safety and health of employees.
(d)(2)(ii)
Where the original technical information no longer exists, such information may be developed in conjunction with the process hazard analysis in sufficient detail to support the analysis.
(d)(3)
Information pertaining to the equipment in the process.
(d)(3)(i)
Information pertaining to the equipment in the process shall include:
(d)(3)(i)(A)
Materials of construction;
(d)(3)(i)(B)
Piping and instrument diagrams (P&ID's);

..1910.119(d)(3)(i)(C)
(d)(3)(i)(C)
Electrical classification;
(d)(3)(i)(D)
Relief system design and design basis;
(d)(3)(i)(E)
Ventilation system design;
(d)(3)(i)(F)
Design codes and standards employed;
(d)(3)(i)(G)
Material and energy balances for processes built after May 26, 1992; and,
(d)(3)(i)(H)
Safety systems (e.g. interlocks, detection or suppression systems).
(d)(3)(ii)
The employer shall document that equipment complies with recognized and generally accepted good engineering practices.
(d)(3)(iii)
For existing equipment designed and constructed in accordance with codes, standards, or practices that are no longer in general use, the employer shall determine and document that the equipment is designed, maintained, inspected, tested, and operating in a safe manner.

..1910.119(e)
(e)
Process hazard analysis.
(e)(1)
The employer shall perform an initial process hazard analysis (hazard evaluation) on processes covered by this standard. The process hazard analysis shall be appropriate to the complexity of the process and shall identify, evaluate, and control the hazards involved in the process. Employers shall determine and document the priority order for conducting process hazard analyses based on a rationale which includes such considerations as extent of the process hazards, number of potentially affected employees, age of the process, and operating history of the process. The process hazard analysis shall be conducted as soon as possible, but not later than the following schedule:
(e)(1)(i)
No less than 25 percent of the initial process hazards analyses shall be completed by May 26, 1994;
(e)(1)(ii)
No less than 50 percent of the initial process hazards analyses shall be completed by May 26, 1995;
(e)(1)(iii)
No less than 75 percent of the initial process hazards analyses shall be completed by May 26, 1996;
(e)(1)(iv)
All initial process hazards analyses shall be completed by May 26, 1997.
(e)(1)(v)
Process hazards analyses completed after May 26, 1987 which meet the requirements of this paragraph are acceptable as initial process hazards analyses. These process hazard analyses shall be updated and revalidated, based on their completion date, in accordance with paragraph (e)(6) of this standard.
(e)(2)
The employer shall use one or more of the following methodologies that are appropriate to determine and evaluate the hazards of the process being analyzed.
(e)(2)(i)
What-If;

..1910.119(e)(2)(ii)
(e)(2)(ii)
Checklist;
(e)(2)(iii)
What-If/Checklist;
(e)(2)(iv)
Hazard and Operability Study (HAZOP);
(e)(2)(v)
Failure Mode and Effects Analysis (FMEA);
(e)(2)(vi)
Fault Tree Analysis; or
(e)(2)(vii)
An appropriate equivalent methodology.
(e)(3)
The process hazard analysis shall address:
(e)(3)(i)
The hazards of the process;
(e)(3)(ii)
The identification of any previous incident which had a likely potential for catastrophic consequences in the workplace;
(e)(3)(iii)
Engineering and administrative controls applicable to the hazards and their interrelationships such as appropriate application of detection

methodologies to provide early warning of releases. (Acceptable detection methods might include process monitoring and control instrumentation with alarms, and detection hardware such as hydrocarbon sensors.);

..1910.119(e)(3)(iv)
(e)(3)(iv)
Consequences of failure of engineering and administrative controls;
(e)(3)(v)
Facility siting;
(e)(3)(vi)
Human factors; and
(e)(3)(vii)
A qualitative evaluation of a range of the possible safety and health effects of failure of controls on employees in the workplace.
(e)(4)
The process hazard analysis shall be performed by a team with expertise in engineering and process operations, and the team shall include at least one employee who has experience and knowledge specific to the process being evaluated. Also, one member of the team must be knowledgeable in the specific process hazard analysis methodology being used.
(e)(5)
The employer shall establish a system to promptly address the team's findings and recommendations; assure that the recommendations are resolved in a timely manner and that the resolution is documented; document what actions are to be taken; complete actions as soon as possible; develop a written schedule of when these actions are to be completed; communicate the actions to operating, maintenance and other employees whose work assignments are in the process and who may be affected by the recommendations or actions.

..1910.119(e)(6)
(e)(6)
At least every five (5) years after the completion of the initial process hazard analysis, the process hazard analysis shall be updated and revalidated by a team meeting the requirements in paragraph (e)(4) of this section, to assure that the process hazard analysis is consistent with the current process.
(e)(7)
Employers shall retain process hazards analyses and updates or revalidations for each process covered by this section, as well as the documented resolution of recommendations described in paragraph (e)(5) of this section for the life of the process.
(f)
Operating procedures.
(f)(1)
The employer shall develop and implement written operating procedures that provide clear instructions for safely conducting activities involved in each covered process consistent with the process safety information and shall address at least the following elements.
(f)(1)(i)
Steps for each operating phase:
(f)(1)(i)(A)
Initial startup;
(f)(1)(i)(B)
Normal operations;
(f)(1)(i)(C)
Temporary operations;

..1910.119(f)(1)(i)(D)
(f)(1)(i)(D)
Emergency shutdown including the conditions under which emergency shutdown is required, and the assignment of shutdown responsibility to qualified operators to ensure that emergency shutdown is executed in a safe and timely manner.
(f)(1)(i)(E)
Emergency Operations;
(f)(1)(i)(F)
Normal shutdown; and,
(f)(1)(i)(G)
Startup following a turnaround, or after an emergency shutdown.
(f)(1)(ii)
Operating limits:
(f)(1)(ii)(A)
Consequences of deviation; and
(f)(1)(ii)(B)
Steps required to correct or avoid deviation.
(f)(1)(iii)
Safety and health considerations:
(f)(1)(iii)(A)
Properties of, and hazards presented by, the chemicals used in the process;
(f)(1)(iii)(B)
Precautions necessary to prevent exposure, including engineering controls, administrative controls, and personal protective equipment;

..1910.119(f)(1)(iii)(C)
(f)(1)(iii)(C)
Control measures to be taken if physical contact or airborne exposure occurs;
(f)(1)(iii)(D)
Quality control for raw materials and control of hazardous chemical inventory levels; and,
(f)(1)(iii)(E)
Any special or unique hazards.
(f)(1)(iv)
Safety systems and their functions.
(f)(2)
Operating procedures shall be readily accessible to employees who work in or maintain a process.
(f)(3)
The operating procedures shall be reviewed as often as necessary to assure that they reflect current operating practice, including changes that result from changes in process chemicals, technology, and equipment, and changes to facilities. The employer shall certify annually that these operating procedures are current and accurate.

..1910.119(f)(4)
(f)(4)
The employer shall develop and implement safe work practices to provide for the control of hazards during operations such as lockout/tagout; confined space entry; opening process equipment or piping; and control over entrance into a facility by maintenance, contractor, laboratory, or other support personnel. These safe work practices shall apply to employees and contractor employees.
(g)
Training.
(g)(1)
Initial training.
(g)(1)(i)
Each employee presently involved in operating a process, and each employee before being involved in operating a newly assigned process, shall be trained in an overview of the process and in the operating procedures as specified in paragraph (f) of this section. The training shall include emphasis on the specific safety and health hazards, emergency operations including shutdown, and safe work practices applicable to the employee's job tasks.
(g)(1)(ii)
In lieu of initial training for those employees already involved in operating a process on May 26, 1992, an employer may certify in writing that the employee has the required knowledge, skills, and abilities to safely carry out the duties and responsibilities as specified in the operating procedures.
(g)(2)
Refresher training. Refresher training shall be provided at least every three years, and more often if necessary, to each employee involved in operating a process to assure that the employee understands and adheres to the current operating procedures of the process. The employer, in consultation with the employees involved in operating the process, shall determine the appropriate frequency of refresher training.

..1910.119(g)(3)
(g)(3)
Training documentation. The employer shall ascertain that each employee involved in operating a process has received and understood the

training required by this paragraph. The employer shall prepare a record which contains the identity of the employee, the date of training, and the means used to verify that the employee understood the training.
(h)
Contractors.
(h)(1)
Application. This paragraph applies to contractors performing maintenance or repair, turnaround, major renovation, or specialty work on or adjacent to a covered process. It does not apply to contractors providing incidental services which do not influence process safety, such as janitorial work, food and drink services, laundry, delivery or other supply services.
(h)(2)
Employer responsibilities.
(h)(2)(i)
The employer, when selecting a contractor, shall obtain and evaluate information regarding the contract employer's safety performance and programs.
(h)(2)(ii)
The employer shall inform contract employers of the known potential fire, explosion, or toxic release hazards related to the contractor's work and the process.
(h)(2)(iii)
The employer shall explain to contract employers the applicable provisions of the emergency action plan required by paragraph (n) of this section.

..1910.119(h)(2)(iv)
(h)(2)(iv)
The employer shall develop and implement safe work practices consistent with paragraph (f)(4) of this section, to control the entrance, presence and exit of contract employers and contract employees in covered process areas.
(h)(2)(v)
The employer shall periodically evaluate the performance of contract employers in fulfilling their obligations as specified in paragraph (h)(3) of this section.
(h)(2)(vi)
The employer shall maintain a contract employee injury and illness log related to the contractor's work in process areas.
(h)(3)
Contract employer responsibilities.
(h)(3)(i)
The contract employer shall assure that each contract employee is trained in the work practices necessary to safely perform his/her job.
(h)(3)(ii)
The contract employer shall assure that each contract employee is instructed in the known potential fire, explosion, or toxic release hazards related to his/her job and the process, and the applicable provisions of the emergency action plan.
(h)(3)(iii)
The contract employer shall document that each contract employee has received and understood the training required by this paragraph. The contract employer shall prepare a record which contains the identity of the contract employee, the date of training, and the means used to verify that the employee understood the training.

..1910.119(h)(3)(iv)
(h)(3)(iv)
The contract employer shall assure that each contract employee follows the safety rules of the facility including the safe work practices required by paragraph (f)(4) of this section.
(h)(3)(v)
The contract employer shall advise the employer of any unique hazards presented by the contract employer's work, or of any hazards found by the contract employer's work.
(i)
Pre-startup safety review.
(i)(1)
The employer shall perform a pre-startup safety review for new facilities and for modified facilities when the modification is significant enough to require a change in the process safety information.
(i)(2)
The pre-startup safety review shall confirm that prior to the introduction of highly hazardous chemicals to a process:
(i)(2)(i)
Construction and equipment is in accordance with design specifications;
(i)(2)(ii)
Safety, operating, maintenance, and emergency procedures are in place and are adequate;
(i)(2)(iii)
For new facilities, a process hazard analysis has been performed and recommendations have been resolved or implemented before startup; and modified facilities meet the requirements contained in management of change, paragraph (l).

..1910.119(i)(2)(iv)
(i)(2)(iv)
Training of each employee involved in operating a process has been completed.
(j)
Mechanical integrity.
(j)(1)
Application. Paragraphs (j)(2) through (j)(6) of this section apply to the following process equipment:
(j)(1)(i)
Pressure vessels and storage tanks;
(j)(1)(ii)
Piping systems (including piping components such as valves);
(j)(1)(iii)
Relief and vent systems and devices;
(j)(1)(iv)
Emergency shutdown systems;
(j)(1)(v)
Controls (including monitoring devices and sensors, alarms, and interlocks) and,
(j)(1)(vi)
Pumps.
(j)(2)
Written procedures. The employer shall establish and implement written procedures to maintain the on-going integrity of process equipment.

..1910.119(j)(3)
(j)(3)
Training for process maintenance activities. The employer shall train each employee involved in maintaining the on-going integrity of process equipment in an overview of that process and its hazards and in the procedures applicable to the employee's job tasks to assure that the employee can perform the job tasks in a safe manner.
(j)(4)
Inspection and testing.
(j)(4)(i)
Inspections and tests shall be performed on process equipment.
(j)(4)(ii)
Inspection and testing procedures shall follow recognized and generally accepted good engineering practices.
(j)(4)(iii)
The frequency of inspections and tests of process equipment shall be consistent with applicable manufacturers' recommendations and good engineering practices, and more frequently if determined to be necessary by prior operating experience.
(j)(4)(iv)
The employer shall document each inspection and test that has been performed on process equipment. The documentation shall identify the date of the inspection or test, the name of the person who performed the inspection or test, the serial number or other identifier of the equipment on which the inspection or test was performed, a description of the inspection or test performed, and the results of the inspection or test.

..1910.119(j)(5)
(j)(5)
Equipment deficiencies. The employer shall correct deficiencies in equipment that are outside acceptable limits (defined by the process safety information in paragraph (d) of this section) before further use or in a safe and timely manner when necessary means are taken to assure safe operation.
(j)(6)
Quality assurance.

(j)(6)(i)
In the construction of new plants and equipment, the employer shall assure that equipment as it is fabricated is suitable for the process application for which they will be used.
(j)(6)(ii)
Appropriate checks and inspections shall be performed to assure that equipment is installed properly and consistent with design specifications and the manufacturer's instructions.
(j)(6)(iii)
The employer shall assure that maintenance materials, spare parts and equipment are suitable for the process application for which they will be used.
(k)
Hot work permit.
(k)(1)
The employer shall issue a hot work permit for hot work operations conducted on or near a covered process.

..1910.119(k)(2)
(k)(2)
The permit shall document that the fire prevention and protection requirements in 29 CFR 1910.252(a) have been implemented prior to beginning the hot work operations; it shall indicate the date(s) authorized for hot work; and identify the object on which hot work is to be performed. The permit shall be kept on file until completion of the hot work operations.
(l)
Management of change.
(l)(1)
The employer shall establish and implement written procedures to manage changes (except for "replacements in kind") to process chemicals, technology, equipment, and procedures; and, changes to facilities that affect a covered process.
(l)(2)
The procedures shall assure that the following considerations are addressed prior to any change:
(l)(2)(i)
The technical basis for the proposed change;
(l)(2)(ii)
Impact of change on safety and health;
(l)(2)(iii)
Modifications to operating procedures;
(l)(2)(iv)
Necessary time period for the change; and,
(l)(2)(v)
Authorization requirements for the proposed change.
(l)(3)
Employees involved in operating a process and maintenance and contract employees whose job tasks will be affected by a change in the process shall be informed of, and trained in, the change prior to start-up of the process or affected part of the process.

..1910.119(l)(4)
(l)(4)
If a change covered by this paragraph results in a change in the process safety information required by paragraph (d) of this section, such information shall be updated accordingly.
(l)(5)
If a change covered by this paragraph results in a change in the operating procedures or practices required by paragraph (f) of this section, such procedures or practices shall be updated accordingly.
(m)
Incident investigation.
(m)(1)
The employer shall investigate each incident which resulted in, or could reasonably have resulted in a catastrophic release of highly hazardous chemical in the workplace.
(m)(2)
An incident investigation shall be initiated as promptly as possible, but not later than 48 hours following the incident.
(m)(3)
An incident investigation team shall be established and consist of at least one person knowledgeable in the process involved, including a contract employee if the incident involved work of the contractor, and other persons with appropriate knowledge and experience to thoroughly investigate and analyze the incident.
(m)(4)
A report shall be prepared at the conclusion of the investigation which includes at a minimum:
(m)(4)(i)
Date of incident;

..1910.119(m)(4)(ii)
(m)(4)(ii)
Date investigation began;
(m)(4)(iii)
A description of the incident;
(m)(4)(iv)
The factors that contributed to the incident; and,
(m)(4)(v)
Any recommendations resulting from the investigation.
(m)(5)
The employer shall establish a system to promptly address and resolve the incident report findings and recommendations. Resolutions and corrective actions shall be documented.
(m)(6)
The report shall be reviewed with all affected personnel whose job tasks are relevant to the incident findings including contract employees where applicable.
(m)(7)
Incident investigation reports shall be retained for five years.

..1910.119(n)
(n)
Emergency planning and response. The employer shall establish and implement an emergency action plan for the entire plant in accordance with the provisions of 29 CFR 1910.38(a). In addition, the emergency action plan shall include procedures for handling small releases. Employers covered under this standard may also be subject to the hazardous waste and emergency response provisions contained in 29 CFR 1910.120(a), (p) and (q).
(o)
Compliance Audits.
(o)(1)
Employers shall certify that they have evaluated compliance with the provisions of this section at least every three years to verify that the procedures and practices developed under the standard are adequate and are being followed.
(o)(2)
The compliance audit shall be conducted by at least one person knowledgeable in the process.
(o)(3)
A report of the findings of the audit shall be developed.
(o)(4)
The employer shall promptly determine and document an appropriate response to each of the findings of the compliance audit, and document that deficiencies have been corrected.
(o)(5)
Employers shall retain the two (2) most recent compliance audit reports.

..1910.119(p)
(p)
Trade secrets.
(p)(1)
Employers shall make all information necessary to comply with the section available to those persons responsible for compiling the process safety information (required by paragraph (d) of this section), those assisting in the development of the process hazard analysis (required by paragraph (e) of this section), those responsible for developing the operating procedures (required by paragraph (f) of this section), and those involved in incident investigations (required by paragraph (m) of this section), emergency planning and response (paragraph (n) of this section) and compliance audits (paragraph (o) of this section) without regard to possible trade secret status of such information.
(p)(2)
Nothing in this paragraph shall preclude the employer from requiring the persons to whom the information is made available under paragraph

(p)(1) of this section to enter into confidentiality agreements not to disclose the information as set forth in 29 CFR 1910.1200.

(p)(3)

Subject to the rules and procedures set forth in 29 CFR 1910.1200(i)(1) through 1910.1200(i)(12), employees and their designated representatives shall have access to trade secret information contained within the process hazard analysis and other documents required to be developed by this standard.

[57 FR 23060, June 1, 1992; 61 FR 9227, March 7, 1996]

Appendix B

OSHA Regulations (Standards - 29 CFR) Hazard Communication. - 1910.1200

Standard Number: 1910.1200
Standard Title: Hazard Communication.
SubPart Number: Z
SubPart Title: Toxic and Hazardous Substances

a)
Purpose."
a)(1)
he purpose of this section is to ensure that the hazards of all chemicals roduced or imported are evaluated, and that information concerning heir hazards is transmitted to employers and employees. This transmital of information is to be accomplished by means of comprehensive azard communication programs, which are to include container labeling nd other forms of warning, material safety data sheets and employee raining.

.1910.1200(a)(2)
a)(2)
his occupational safety and health standard is intended to address omprehensively the issue of evaluating the potential hazards of chemials, and communicating information concerning hazards and appropriate protective measures to employees, and to preempt any legal requirenents of a state, or political subdivision of a state, pertaining to this subect. Evaluating the potential hazards of chemicals, and communicating nformation concerning hazards and appropriate protective measures to mployees, may include, for example, but is not limited to, provisions for: leveloping and maintaining a written hazard communication program for he workplace, including lists of hazardous chemicals present; labeling of ontainers of chemicals in the workplace, as well as of containers of hemicals being shipped to other workplaces; preparation and distribuion of material safety data sheets to employees and downstream mployers; and development and implementation of employee training rograms regarding hazards of chemicals and protective measures. Jnder section 18 of the Act, no state or political subdivision of a state nay adopt or enforce, through any court or agency, any requirement elating to the issue addressed by this Federal standard, except pursuant to a Federally-approved state plan.
b)
Scope and application."
b)(1)
his section requires chemical manufacturers or importers to assess the azards of chemicals which they produce or import, and all employers to rovide information to their employees about the hazardous chemicals to vhich they are exposed, by means of a hazard communication program, abels and other forms of warning, material safety data sheets, and infornation
nd training. In addition, this section requires distributors to transmit the equired information to employers. (Employers who do not produce or nport chemicals need only focus on those parts of this rule that deal vith establishing a workplace program and communicating information to heir workers. Appendix E of this section is a general guide for such mployers to help them determine their compliance obligations under the ule.)
b)(2)
his section applies to any chemical which is known to be present in the vorkplace in such a manner that employees may be exposed under noral conditions of use or in a foreseeable emergency.

(b)(3)
This section applies to laboratories only as follows:
(b)(3)(i)
Employers shall ensure that labels on incoming containers of hazardous chemicals are not removed or defaced;

..1910.1200(b)(3)(ii)
(b)(3)(ii)
Employers shall maintain any material safety data sheets that are received with incoming shipments of hazardous chemicals, and ensure that they are readily accessible during each workshift to laboratory employees when they are in their work areas;
(b)(3)(iii)
Employers shall ensure that laboratory employees are provided information and training in accordance with paragraph (h) of this section, except for the location and availability of the written hazard communication program under paragraph (h)(2)(iii) of this section; and,
(b)(3)(iv)
Laboratory employers that ship hazardous chemicals are considered to be either a chemical manufacturer or a distributor under this rule, and thus must ensure that any containers of hazardous chemicals leaving the laboratory are labeled in accordance with paragraph (f)(1) of this section, and that a material safety data sheet is provided to distributors and other employers in accordance with paragraphs (g)(6) and (g)(7) of this section.
(b)(4)
In work operations where employees only handle chemicals in sealed containers which are not opened under normal conditions of use (such as are found in marine cargo handling, warehousing, or retail sales), this section applies to these operations only as follows:
(b)(4)(i)
Employers shall ensure that labels on incoming containers of hazardous chemicals are not removed or defaced;

..1910.1200(b)(4)(ii)
(b)(4)(ii)
Employers shall maintain copies of any material safety data sheets that are received with incoming shipments of the sealed containers of hazardous chemicals, shall obtain a material safety data sheet as soon as possible for sealed containers of hazardous chemicals received without a material safety data sheet if an employee requests the material safety data sheet, and shall ensure that the material safety data sheets are readily accessible during each work shift to employees when they are in their work area(s); and,
(b)(4)(iii)
Employers shall ensure that employees are provided with information and training in accordance with paragraph (h) of this section (except for the location and availability of the written hazard communication program under paragraph (h)(2)(iii) of this section), to the extent necessary to protect them in the event of a spill or leak of a hazardous chemical from a sealed container.

(b)(5)
This section does not require labeling of the following chemicals:
(b)(5)(i)
Any pesticide as such term is defined in the Federal Insecticide, Fungicide, and Rodenticide Act (7 U.S.C. 136 et seq.), when subject to the labeling requirements of that Act and labeling regulations issued under that Act by the Environmental Protection Agency;
(b)(5)(ii)
Any chemical substance or mixture as such terms are defined in the Toxic Substances Control Act (15 U.S.C. 2601 et seq.), when subject to the labeling requirements of that Act and labeling regulations issued under that Act by the Environmental Protection Agency;

..1910.1200(b)(5)(iii)
(b)(5)(iii)
Any food, food additive, color additive, drug, cosmetic, or medical or veterinary device or product, including materials intended for use as ingredients in such products (e.g. flavors and fragrances), as such terms are defined in the Federal Food, Drug, and Cosmetic Act (21 U.S.C. 301 et seq.) or the Virus-Serum-Toxin Act of 1913 (21 U.S.C. 151 et seq.), and regulations issued under those Acts, when they are subject to the labeling requirements under those Acts by either the Food and Drug Administration or the Department of Agriculture;
(b)(5)(iv)
Any distilled spirits (beverage alcohols), wine, or malt beverage intended for nonindustrial use, as such terms are defined in the Federal Alcohol Administration Act (27 U.S.C. 201 et seq.) and regulations issued under that Act, when subject to the labeling requirements of that Act and labeling regulations issued under that Act by the Bureau of Alcohol, Tobacco, and Firearms;
(b)(5)(v)
Any consumer product or hazardous substance as those terms are defined in the Consumer Product Safety Act (15 U.S.C. 2051 et seq.) and Federal Hazardous Substances Act (15 U.S.C. 1261 et seq.) respectively, when subject to a consumer product safety standard or labeling requirement of those Acts, or regulations issued under those Acts by the Consumer Product Safety Commission; and,
(b)(5)(vi)
Agricultural or vegetable seed treated with pesticides and labeled in accordance with the Federal Seed Act (7 U.S.C. 1551 et seq.) and the labeling regulations issued under that Act by the Department of Agriculture.

..1910.1200(b)(6)
(b)(6)
This section does not apply to:
(b)(6)(i)
Any hazardous waste as such term is defined by the Solid Waste Disposal Act, as amended by the Resource Conservation and Recovery Act of 1976, as amended (42 U.S.C. 6901 et seq.), when subject to regulations issued under that Act by the Environmental Protection Agency;
(b)(6)(ii)
Any hazardous substance as such term is defined by the Comprehensive Environmental Response, Compensation and Liability ACT (CERCLA) (42 U.S.C. 9601 et seq.) when the hazardous substance is the focus of remedial or removal action being conducted under CERCLA in accordance with the Environmental Protection Agency regulations.
(b)(6)(iii)
Tobacco or tobacco products;
(b)(6)(iv)
Wood or wood products, including lumber which will not be processed, where the chemical manufacturer or importer can establish that the only hazard they pose to employees is the potential for flammability or combustibility (wood or wood products which have been treated with a hazardous chemical covered by this standard, and wood which may be subsequently sawed or cut, generating dust, are not exempted);
(b)(6)(v)
Articles (as that term is defined in paragraph (c) of this section);
(b)(6)(vi)
Food or alcoholic beverages which are sold, used, or prepared in a retail establishment (such as a grocery store, restaurant, or drinking place), and foods intended for personal consumption by employees while in the workplace;

..1910.1200(b)(6)(vii)
(b)(6)(vii)
Any drug, as that term is defined in the Federal Food, Drug, and Cosmetic Act (21 U.S.C. 301 et seq.), when it is in solid, final form for direct administration to the patient (e.g., tablets or pills); drugs which are packaged by the chemical manufacturer for sale to consumers in a retail establishment (e.g., over-the-counter drugs); and drugs intended for personal consumption by employees while in the workplace (e.g., first aid supplies);
(b)(6)(viii)
Cosmetics which are packaged for sale to consumers in a retail establishment, and cosmetics intended for personal consumption by employees while in the workplace;
(b)(6)(ix)
Any consumer product or hazardous substance, as those terms are defined in the Consumer Product Safety Act (15 U.S.C. 2051 et seq.) and Federal Hazardous Substances Act (15 U.S.C. 1261 et seq.) respectively, where the employer can show that it is used in the workplace for the purpose intended by the chemical manufacturer or importer of the product, and the use results in a duration and frequency of exposure which is not greater than the range of exposures that could reasonably be experienced by consumers when used for the purpose intended;
(b)(6)(x)
Nuisance particulates where the chemical manufacturer or importer can establish that they do not pose any physical or health hazard covered under this section;
(b)(6)(xi)
Ionizing and nonionizing radiation; and,
(b)(6)(xii)
Biological hazards.
(c)
"Definitions."
"Article" means a manufactured item other than a fluid or particle:
(i) which is formed to a specific shape or design during manufacture;
(ii) which has end use function(s) dependent in whole or in part upon its shape or design during end use; and
(iii) which under normal conditions of use does not release more than very small quantities, e.g., minute or trace amounts of a hazardous chemical (as determined under paragraph (d) of this section), and does not pose a physical hazard or health risk to employees.
"Assistant Secretary" means the Assistant Secretary of Labor for Occupational Safety and Health, U.S. Department of Labor, or designee.
"Chemical" means any element, chemical compound or mixture of elements and/or compounds.
"Chemical manufacturer" means an employer with a workplace where chemical(s) are produced for use or distribution.
"Chemical name" means the scientific designation of a chemical in accordance with the nomenclature system developed by the International Union of Pure and Applied Chemistry (IUPAC) or the Chemical Abstracts Service (CAS) rules of nomenclature, or a name which will clearly identify the chemical for the purpose of conducting a hazard evaluation.
"Combustible liquid" means any liquid having a flashpoint at or above 100 deg. F (37.8 deg. C), but below 200 deg. F (93.3 deg. C), except any mixture having components with flashpoints of 200 deg. F (93.3 deg. C), or higher, the total volume of which make up 99 percent or more of the total volume of the mixture.
"Commercial account" means an arrangement whereby a retail distributor sells hazardous chemicals to an employer, generally in large quantities over time and/or at costs that are below the regular retail price.
"Common name" means any designation or identification such as code name, code number, trade name, brand name or generic name used to identify a chemical other than by its chemical name.
"Compressed gas" means:
(i) A gas or mixture of gases having, in a container, an absolute pressure exceeding 40 psi at 70 deg. F (21.1 deg. C); or
(ii) A gas or mixture of gases having, in a container, an absolute pressure exceeding 104 psi at 130 deg. F (54.4 deg. C) regardless of the pressure at 70 deg. F (21.1 deg. C); or
(iii) A liquid having a vapor pressure exceeding 40 psi at 100 deg. F (37.8 deg. C) as determined by ASTM D-323-72.
"Container" means any bag, barrel, bottle, box, can, cylinder, drum, reaction vessel, storage tank, or the like that contains a hazardous chemical.

For purposes of this section, pipes or piping systems, and engines, fuel tanks, or other operating systems in a vehicle, are not considered to be containers.
"Designated representative" means any individual or organization to whom an employee gives written authorization to exercise such employee's rights under this section. A recognized or certified collective bargaining agent shall be treated automatically as a designated representative without regard to written employee authorization.
"Director" means the Director, National Institute for Occupational Safety and Health, U.S. Department of Health and Human Services, or designee.
"Distributor" means a business, other than a chemical manufacturer or importer, which supplies hazardous chemicals to other distributors or to employers.
"Employee" means a worker who may be exposed to hazardous chemicals under normal operating conditions or in foreseeable emergencies. Workers such as office workers or bank tellers who encounter hazardous chemicals only in non-routine, isolated instances are not covered.
"Employer" means a person engaged in a business where chemicals are either used, distributed, or are produced for use or distribution, including a contractor or subcontractor.
"Explosive" means a chemical that causes a sudden, almost instantaneous release of pressure, gas, and heat when subjected to sudden shock, pressure, or high temperature.
"Exposure or exposed" means that an employee is subjected in the course of employment to a chemical that is a physical or health hazard, and includes potential (e.g. accidental or possible) exposure. "Subjected" in terms of health hazards includes any route of entry (e.g. inhalation, ingestion, skin contact or absorption.)
"Flammable" means a chemical that falls into one of the following categories:
(i) "Aerosol, flammable" means an aerosol that, when tested by the method described in 16 CFR 1500.45, yields a flame projection exceeding 18 inches at full valve opening, or a flashback (a flame extending back to the valve) at any degree of valve opening;
(ii) "Gas, flammable" means: (A) A gas that, at ambient temperature and pressure, forms a flammable mixture with air at a concentration of thirteen (13) percent by volume or less; or (B) A gas that, at ambient temperature and pressure, forms a range of flammable mixtures with air wider than twelve (12) percent by volume, regardless of the lower limit;
(iii) "Liquid, flammable" means any liquid having a flashpoint below 100 deg. F (37.8 deg. C), except any mixture having components with flashpoints of 100 deg. F (37.8 deg. C) or higher, the total of which make up 99 percent or more of the total volume of the mixture.
(iv) "Solid, flammable" means a solid, other than a blasting agent or explosive as defined in 1910.109(a), that is liable to cause fire through friction, absorption of moisture, spontaneous chemical change, or retained heat from manufacturing or processing, or which can be ignited readily and when ignited burns so vigorously and persistently as to create a serious hazard. A chemical shall be considered to be a flammable solid if, when tested by the method described in 16 CFR 1500.44, it ignites and burns with a self-sustained flame at a rate greater than one-tenth of an inch per second along its major axis.
"Flashpoint" means the minimum temperature at which a liquid gives off a vapor in sufficient concentration to ignite when tested as follows:
(i) Tagliabue Closed Tester (See American National Standard Method of Test for Flash Point by Tag Closed Tester, Z11.24-1979 (ASTM D 56-79)) for liquids with a viscosity of less than 45 Saybolt Universal Seconds (SUS) at 100 deg. F (37.8 deg. C), that do not contain suspended solids and do not have a tendency to form a surface film under test; or
(ii) Pensky-Martens Closed Tester (see American National Standard Method of Test for Flash Point by Pensky-Martens Closed Tester, Z11.7-1979 (ASTM D 93-79)) for liquids with a viscosity equal to or greater than 45 SUS at 100 deg. F (37.8 deg. C), or that contain suspended solids, or that have a tendency to form a surface film under test; or
(iii) Setaflash Closed Tester (see American National Standard Method of Test for Flash Point by Setaflash Closed Tester (ASTM D 3278-78)).
Organic peroxides, which undergo autoaccelerating thermal decomposition, are excluded from any of the flashpoint determination methods specified above.
"Foreseeable emergency" means any potential occurrence such as, but not limited to, equipment failure, rupture of containers, or failure of control equipment which could result in an uncontrolled release of a hazardous chemical into the workplace.
"Hazardous chemical" means any chemical which is a physical hazard or a health hazard.
"Hazard warning" means any words, pictures, symbols, or combination thereof appearing on a label or other appropriate form of warning which convey the specific physical and health hazard(s), including target organ effects, of the chemical(s) in the container(s). (See the definitions for "physical hazard" and "health hazard" to determine the hazards which must be covered.)
"Health hazard" means a chemical for which there is statistically significant evidence based on at least one study conducted in accordance with established scientific principles that acute or chronic health effects may occur in exposed employees. The term "health hazard" includes chemicals which are carcinogens, toxic or highly toxic agents, reproductive toxins, irritants, corrosives, sensitizers, hepatotoxins, nephrotoxins, neurotoxins, agents which act on the hematopoietic system, and agents which damage the lungs, skin, eyes, or mucous membranes. Appendix A provides further definitions and explanations of the scope of health hazards covered by this section, and Appendix B describes the criteria to be used to determine whether or not a chemical is to be considered hazardous for purposes of this standard.
"Identity" means any chemical or common name which is indicated on the material safety data sheet (MSDS) for the chemical. The identity used shall permit cross-references to be made among the required list of hazardous chemicals, the label and the MSDS.
"Immediate use" means that the hazardous chemical will be under the control of and used only by the person who transfers it from a labeled container and only within the work shift in which it is transferred.
"Importer" means the first business with employees within the Customs Territory of the United States which receives hazardous chemicals produced in other countries for the purpose of supplying them to distributors or employers within the United States.
"Label" means any written, printed, or graphic material displayed on or affixed to containers of hazardous chemicals.
"Material safety data sheet (MSDS)" means written or printed material concerning a hazardous chemical which is prepared in accordance with paragraph (g) of this section.
"Mixture" means any combination of two or more chemicals if the combination is not, in whole or in part, the result of a chemical reaction.
"Organic peroxide" means an organic compound that contains the bivalent -O-O-structure and which may be considered to be a structural derivative of hydrogen peroxide where one or both of the hydrogen atoms has been replaced by an organic radical.
"Oxidizer" means a chemical other than a blasting agent or explosive as defined in 1910.109(a), that initiates or promotes combustion in other materials, thereby causing fire either of itself or through the release of oxygen or other gases.
"Physical hazard" means a chemical for which there is scientifically valid evidence that it is a combustible liquid, a compressed gas, explosive, flammable, an organic peroxide, an oxidizer, pyrophoric, unstable (reactive) or water-reactive.
"Produce" means to manufacture, process, formulate, blend, extract, generate, emit, or repackage.
"Pyrophoric" means a chemical that will ignite spontaneously in air at a temperature of 130 deg. F (54.4 deg. C) or below.
"Responsible party" means someone who can provide additional information on the hazardous chemical and appropriate emergency procedures, if necessary.
"Specific chemical identity" means the chemical name, Chemical Abstracts Service (CAS) Registry Number, or any other information that reveals the precise chemical designation of the substance.
"Trade secret" means any confidential formula, pattern, process, device, information or compilation of information that is used in an employer's business, and that gives the employer an opportunity to obtain an advantage over competitors who do not know or use it. Appendix D sets out the criteria to be used in evaluating trade secrets.
"Unstable (reactive)" means a chemical which in the pure state, or as produced or transported, will vigorously polymerize, decompose, condense, or will become self-reactive under conditions of shocks, pressure or temperature.
"Use" means to package, handle, react, emit, extract, generate as a byproduct, or transfer.
"Water-reactive" means a chemical that reacts with water to release a gas that is either flammable or presents a health hazard.
"Work area" means a room or defined space in a workplace where haz-

ardous chemicals are produced or used, and where employees are present.
"Workplace" means an establishment, job site, or project, at one geographical location containing one or more work areas.

..1910.1200(d)
(d)
"Hazard determination."
(d)(1)
Chemical manufacturers and importers shall evaluate chemicals produced in their workplaces or imported by them to determine if they are hazardous. Employers are not required to evaluate chemicals unless they choose not to rely on the evaluation performed by the chemical manufacturer or importer for the chemical to satisfy this requirement.
(d)(2)
Chemical manufacturers, importers or employers evaluating chemicals shall identify and consider the available scientific evidence concerning such hazards. For health hazards, evidence which is statistically significant and which is based on at least one positive study conducted in accordance with established scientific principles is considered to be sufficient to establish a hazardous effect if the results of the study meet the definitions of health hazards in this section. Appendix A shall be consulted for the scope of health hazards covered, and Appendix B shall be consulted for the criteria to be followed with respect to the completeness of the evaluation, and the data to be reported.
(d)(3)
The chemical manufacturer, importer or employer evaluating chemicals shall treat the following sources as establishing that the chemicals listed in them are hazardous:
(d)(3)(i)
29 CFR part 1910, subpart Z, Toxic and Hazardous Substances, Occupational Safety and Health Administration (OSHA); or,

..1910.1200(d)(3)(ii)
(d)(3)(ii)
"Threshold Limit Values for Chemical Substances and Physical Agents in the Work Environment," American Conference of Governmental Industrial Hygienists (ACGIH) (latest edition). The chemical manufacturer, importer, or employer is still responsible for evaluating the hazards associated with the chemicals in these source lists in accordance with the requirements of this standard.
(d)(4)
Chemical manufacturers, importers and employers evaluating chemicals shall treat the following sources as establishing that a chemical is a carcinogen or potential carcinogen for hazard communication purposes:
(d)(4)(i)
National Toxicology Program (NTP), "Annual Report on Carcinogens" (latest edition);
(d)(4)(ii)
International Agency for Research on Cancer (IARC) "Monographs" (latest editions); or
(d)(4)(iii)
29 CFR part 1910, subpart Z, Toxic and Hazardous Substances, Occupational Safety and Health Administration.

Note: The "Registry of Toxic Effects of Chemical Substances" published by the National Institute for Occupational Safety and Health indicates whether a chemical has been found by NTP or IARC to be a potential carcinogen.
(d)(5)
The chemical manufacturer, importer or employer shall determine the hazards of mixtures of chemicals as follows:
(d)(5)(i)
If a mixture has been tested as a whole to determine its hazards, the results of such testing shall be used to determine whether the mixture is hazardous;

..1910.1200(d)(5)(ii)
(d)(5)(ii)
If a mixture has not been tested as a whole to determine whether the mixture is a health hazard, the mixture shall be assumed to present the same health hazards as do the components which comprise one percent (by weight or volume) or greater of the mixture, except that the mixture shall be assumed to present a carcinogenic hazard if it contains a component in concentrations of 0.1 percent or greater which is considered to be a carcinogen under paragraph (d)(4) of this section;
(d)(5)(iii)
If a mixture has not been tested as a whole to determine whether the mixture is a physical hazard, the chemical manufacturer, importer, or employer may use whatever scientifically valid data is available to evaluate the physical hazard potential of the mixture; and,
(d)(5)(iv)
If the chemical manufacturer, importer, or employer has evidence to indicate that a component present in the mixture in concentrations of less than one percent (or in the case of carcinogens, less than 0.1 percent) could be released in concentrations which would exceed an established OSHA permissible exposure limit or ACGIH Threshold Limit Value, or could present a health risk to employees in those concentrations, the mixture shall be assumed to present the same hazard.
(d)(6)
Chemical manufacturers, importers, or employers evaluating chemicals shall describe in writing the procedures they use to determine the hazards of the chemical they evaluate. The written procedures are to be made available, upon request, to employees, their designated representatives, the Assistant Secretary and the Director. The written description may be incorporated into the written hazard communication program required under paragraph (e) of this section.

..1910.1200(e)
(e)
"Written hazard communication program."
(e)(1)
Employers shall develop, implement, and maintain at each workplace, a written hazard communication program which at least describes how the criteria specified in paragraphs (f), (g), and (h) of this section for labels and other forms of warning, material safety data sheets, and employee information and training will be met, and which also includes the following:
(e)(1)(i)
A list of the hazardous chemicals known to be present using an identity that is referenced on the appropriate material safety data sheet (the list may be compiled for the workplace as a whole or for individual work areas); and,
(e)(1)(ii)
The methods the employer will use to inform employees of the hazards of non-routine tasks (for example, the cleaning of reactor vessels), and the hazards associated with chemicals contained in unlabeled pipes in their work areas.
(e)(2)
"Multi-employer workplaces." Employers who produce, use, or store hazardous chemicals at a workplace in such a way that the employees of other employer(s) may be exposed (for example, employees of a construction contractor working on-site) shall additionally ensure that the hazard communication programs developed and implemented under this paragraph (e) include the following:
(e)(2)(i)
The methods the employer will use to provide the other employer(s) on-site access to material safety data sheets for each hazardous chemical the other employer(s)' employees may be exposed to while working;

..1910.1200(e)(2)(ii)
(e)(2)(ii)
The methods the employer will use to inform the other employer(s) of any precautionary measures that need to be taken to protect employees during the workplace's normal operating conditions and in foreseeable emergencies; and,
(e)(2)(iii)
The methods the employer will use to inform the other employer(s) of the labeling system used in the workplace.
(e)(3)
The employer may rely on an existing hazard communication program to comply with these requirements, provided that it meets the criteria established in this paragraph (e).
(e)(4)
The employer shall make the written hazard communication program available, upon request, to employees, their designated representatives, the Assistant Secretary and the Director, in accordance with the requirements of 29 CFR 1910.1020 (e).

(e)(5)
Where employees must travel between workplaces during a workshift, i.e., their work is carried out at more than one geographical location, the written hazard communication program may be kept at the primary workplace facility.
(f)
"Labels and other forms of warning."
(f)(1)
The chemical manufacturer, importer, or distributor shall ensure that each container of hazardous chemicals leaving the workplace is labeled, tagged or marked with the following information:

..1910.1200(f)(1)(i)
(f)(1)(i)
Identity of the hazardous chemical(s);
(f)(1)(ii)
Appropriate hazard warnings; and
(f)(1)(iii)
Name and address of the chemical manufacturer, importer, or other responsible party.
(f)(2)
(f)(2)(i)
For solid metal (such as a steel beam or a metal casting), solid wood, or plastic items that are not exempted as articles due to their downstream use, or shipments of whole grain, the required label may be transmitted to the customer at the time of the initial shipment, and need not be included with subsequent shipments to the same employer unless the information on the label changes;
(f)(2)(ii)
The label may be transmitted with the initial shipment itself, or with the material safety data sheet that is to be provided prior to or at the time of the first shipment; and,
(f)(2)(iii)
This exception to requiring labels on every container of hazardous chemicals is only for the solid material itself, and does not apply to hazardous chemicals used in conjunction with, or known to be present with, the material and to which employees handling the items in transit may be exposed (for example, cutting fluids or pesticides in grains).

..1910.1200(f)(3)
(f)(3)
Chemical manufacturers, importers, or distributors shall ensure that each container of hazardous chemicals leaving the workplace is labeled, tagged, or marked in accordance with this section in a manner which does not conflict with the requirements of the Hazardous Materials Transportation Act (49 U.S.C. 1801 et seq.) and regulations issued under that Act by the Department of Transportation.
(f)(4)
If the hazardous chemical is regulated by OSHA in a substance-specific health standard, the chemical manufacturer, importer, distributor or employer shall ensure that the labels or other forms of warning used are in accordance with the requirements of that standard.
(f)(5)
Except as provided in paragraphs (f)(6) and (f)(7) of this section, the employer shall ensure that each container of hazardous chemicals in the workplace is labeled, tagged or marked with the following information:
(f)(5)(i)
Identity of the hazardous chemical(s) contained therein; and,
(f)(5)(ii)
Appropriate hazard warnings, or alternatively, words, pictures, symbols, or combination thereof, which provide at least general information regarding the hazards of the chemicals, and which, in conjunction with the other information immediately available to employees under the hazard communication program, will provide employees with the specific information regarding the physical and health hazards of the hazardous chemical.

..1910.1200(f)(6)
(f)(6)
The employer may use signs, placards, process sheets, batch tickets, operating procedures, or other such written materials in lieu of affixing labels to individual stationary process containers, as long as the alternative method identifies the containers to which it is applicable and conveys the information required by paragraph (f)(5) of this section to be on a label. The written materials shall be readily accessible to the employees in their work area throughout each work shift.
(f)(7)
The employer is not required to label portable containers into which hazardous chemicals are transferred from labeled containers, and which are intended only for the immediate use of the employee who performs the transfer. For purposes of this section, drugs which are dispensed by a pharmacy to a health care provider for direct administration to a patient are exempted from labeling.
(f)(8)
The employer shall not remove or deface existing labels on incoming containers of hazardous chemicals, unless the container is immediately marked with the required information.
(f)(9)
The employer shall ensure that labels or other forms of warning are legible, in English, and prominently displayed on the container, or readily available in the work area throughout each work shift. Employers having employees who speak other languages may add the information in their language to the material presented, as long as the information is presented in English as well.
(f)(10)
The chemical manufacturer, importer, distributor or employer need not affix new labels to comply with this section if existing labels already convey the required information.

..1910.1200(f)(11)
(f)(11)
Chemical manufacturers, importers, distributors, or employers who become newly aware of any significant information regarding the hazards of a chemical shall revise the labels for the chemical within three months of becoming aware of the new information. Labels on containers of hazardous chemicals shipped after that time shall contain the new information. If the chemical is not currently produced or imported, the chemical manufacturer, importers, distributor, or employer shall add the information to the label before the chemical is shipped or introduced into the workplace again.
(g)
"Material safety data sheets."
(g)(1)
Chemical manufacturers and importers shall obtain or develop a material safety data sheet for each hazardous chemical they produce or import. Employers shall have a material safety data sheet in the workplace for each hazardous chemical which they use.
(g)(2)
Each material safety data sheet shall be in English (although the employer may maintain copies in other languages as well), and shall contain at least the following information:
(g)(2)(i)
The identity used on the label, and, except as provided for in paragraph (i) of this section on trade secrets:
(g)(2)(i)(A)
If the hazardous chemical is a single substance, its chemical and common name(s);
(g)(2)(i)(B)
If the hazardous chemical is a mixture which has been tested as a whole to determine its hazards, the chemical and common name(s) of the ingredients which contribute to these known hazards, and the common name(s) of the mixture itself; or,
(g)(2)(i)(C)
If the hazardous chemical is a mixture which has not been tested as a whole:

..1910.1200(g)(2)(i)(C)(1)
(g)(2)(i)(C)(1)
The chemical and common name(s) of all ingredients which have been determined to be health hazards, and which comprise 1% or greater of the composition, except that chemicals identified as carcinogens under paragraph (d) of this section shall be listed if the concentrations are 0.1% or greater; and,
(g)(2)(i)(C)(2)
The chemical and common name(s) of all ingredients which have been determined to be health hazards, and which comprise less than 1% (0.1% for carcinogens) of the mixture, if there is evidence that the ingredient(s) could be released from the mixture in concentrations which

would exceed an established OSHA permissible exposure limit or ACGIH Threshold Limit Value, or could present a health risk to employees; and,
(g)(2)(i)(C)(3)
The chemical and common name(s) of all ingredients which have been determined to present a physical hazard when present in the mixture;
(g)(2)(ii)
Physical and chemical characteristics of the hazardous chemical (such as vapor pressure, flash point);
(g)(2)(iii)
The physical hazards of the hazardous chemical, including the potential for fire, explosion, and reactivity;
(g)(2)(iv)
The health hazards of the hazardous chemical, including signs and symptoms of exposure, and any medical conditions which are generally recognized as being aggravated by exposure to the chemical;
(g)(2)(v)
The primary route(s) of entry;

..1910.1200(g)(2)(vi)
(g)(2)(vi)
The OSHA permissible exposure limit, ACGIH Threshold Limit Value, and any other exposure limit used or recommended by the chemical manufacturer, importer, or employer preparing the material safety data sheet, where available;
(g)(2)(vii)
Whether the hazardous chemical is listed in the National Toxicology Program (NTP) Annual Report on Carcinogens (latest edition) or has been found to be a potential carcinogen in the International Agency for Research on Cancer (IARC) Monographs (latest editions), or by OSHA;
(g)(2)(viii)
Any generally applicable precautions for safe handling and use which are known to the chemical manufacturer, importer or employer preparing the material safety data sheet, including appropriate hygienic practices, protective measures during repair and maintenance of contaminated equipment, and procedures for clean-up of spills and leaks;
(g)(2)(ix)
Any generally applicable control measures which are known to the chemical manufacturer, importer or employer preparing the material safety data sheet, such as appropriate engineering controls, work practices, or personal protective equipment;
(g)(2)(x)
Emergency and first aid procedures;
(g)(2)(xi)
The date of preparation of the material safety data sheet or the last change to it; and,

..1910.1200(g)(2)(xii)
(g)(2)(xii)
The name, address and telephone number of the chemical manufacturer, importer, employer or other responsible party preparing or distributing the material safety data sheet, who can provide additional information on the hazardous chemical and appropriate emergency procedures, if necessary.
(g)(3)
If no relevant information is found for any given category on the material safety data sheet, the chemical manufacturer, importer or employer preparing the material safety data sheet shall mark it to indicate that no applicable information was found.
(g)(4)
Where complex mixtures have similar hazards and contents (i.e. the chemical ingredients are essentially the same, but the specific composition varies from mixture to mixture), the chemical manufacturer, importer or employer may prepare one material safety data sheet to apply to all of these similar mixtures.
(g)(5)
The chemical manufacturer, importer or employer preparing the material safety data sheet shall ensure that the information recorded accurately reflects the scientific evidence used in making the hazard determination. If the chemical manufacturer, importer or employer preparing the material safety data sheet becomes newly aware of any significant information regarding the hazards of a chemical, or ways to protect against the hazards, this new information shall be added to the material safety data sheet within three months. If the chemical is not currently being produced or imported the chemical manufacturer or importer shall add the information to the material safety data sheet before the chemical is introduced into the workplace again.

..1910.1200(g)(6)
(g)(6)
(g)(6)(i)
Chemical manufacturers or importers shall ensure that distributors and employers are provided an appropriate material safety data sheet with their initial shipment, and with the first shipment after a material safety data sheet is updated;
(g)(6)(ii)
The chemical manufacturer or importer shall either provide material safety data sheets with the shipped containers or send them to the distributor or employer prior to or at the time of the shipment;
(g)(6)(iii)
If the material safety data sheet is not provided with a shipment that has been labeled as a hazardous chemical, the distributor or employer shall obtain one from the chemical manufacturer or importer as soon as possible; and,
(g)(6)(iv)
The chemical manufacturer or importer shall also provide distributors or employers with a material safety data sheet upon request.
(g)(7)
(g)(7)(i)
Distributors shall ensure that material safety data sheets, and updated information, are provided to other distributors and employers with their initial shipment and with the first shipment after a material safety data sheet is updated;
(g)(7)(ii)
The distributor shall either provide material safety data sheets with the shipped containers, or send them to the other distributor or employer prior to or at the time of the shipment;

..1910.1200(g)(7)(iii)
(g)(7)(iii)
Retail distributors selling hazardous chemicals to employers having a commercial account shall provide a material safety data sheet to such employers upon request, and shall post a sign or otherwise inform them that a material safety data sheet is available;
(g)(7)(iv)
Wholesale distributors selling hazardous chemicals to employers over-the-counter may also provide material safety data sheets upon the request of the employer at the time of the over-the-counter purchase, and shall post a sign or otherwise inform such employers that a material safety data sheet is available;
(g)(7)(v)
If an employer without a commercial account purchases a hazardous chemical from a retail distributor not required to have material safety data sheets on file (i.e., the retail distributor does not have commercial accounts and does not use the materials), the retail distributor shall provide the employer, upon request, with the name, address, and telephone number of the chemical manufacturer, importer, or distributor from which a material safety data sheet can be obtained;
(g)(7)(vi)
Wholesale distributors shall also provide material safety data sheets to employers or other distributors upon request; and,
(g)(7)(vii)
Chemical manufacturers, importers, and distributors need not provide material safety data sheets to retail distributors that have informed them that the retail distributor does not sell the product to commercial accounts or open the sealed container to use it in their own workplaces.

..1910.1200(g)(8)
(g)(8)
The employer shall maintain in the workplace copies of the required material safety data sheets for each hazardous chemical, and shall ensure that they are readily accessible during each work shift to employees when they are in their work area(s). (Electronic access, microfiche, and other alternatives to maintaining paper copies of the material safety data sheets are permitted as long as no barriers to immediate employee access in each workplace are created by such options.)

(g)(9)
Where employees must travel between workplaces during a workshift, i.e., their work is carried out at more than one geographical location, the material safety data sheets may be kept at the primary workplace facility. In this situation, the employer shall ensure that employees can immediately obtain the required information in an emergency.
(g)(10)
Material safety data sheets may be kept in any form, including operating procedures, and may be designed to cover groups of hazardous chemicals in a work area where it may be more appropriate to address the hazards of a process rather than individual hazardous chemicals. However, the employer shall ensure that in all cases the required information is provided for each hazardous chemical, and is readily accessible during each work shift to employees when they are in in their work area(s).
(g)(11)
Material safety data sheets shall also be made readily available, upon request, to designated representatives and to the Assistant Secretary, in accordance with the requirements of 29 CFR 1910.1020(e). The Director shall also be given access to material safety data sheets in the same manner.

..1910.1200(h)
(h)
"Employee information and training."
(h)(1)
Employers shall provide employees with effective information and training on hazardous chemicals in their work area at the time of their initial assignment, and whenever a new physical or health hazard the employees have not previously been trained about is introduced into their work area. Information and training may be designed to cover categories of hazards (e.g., flammability, carcinogenicity) or specific chemicals. Chemical-specific information must always be available through labels and material safety data sheets.
(h)(2)
"Information." Employees shall be informed of:
(h)(2)(i)
The requirements of this section;
(h)(2)(ii)
Any operations in their work area where hazardous chemicals are present; and,
(h)(2)(iii)
The location and availability of the written hazard communication program, including the required list(s) of hazardous chemicals, and material safety data sheets required by this section.
(h)(3)
"Training." Employee training shall include at least:
(h)(3)(i)
Methods and observations that may be used to detect the presence or release of a hazardous chemical in the work area (such as monitoring conducted by the employer, continuous monitoring devices, visual appearance or odor of hazardous chemicals when being released, etc.);
(h)(3)(ii)
The physical and health hazards of the chemicals in the work area;

..1910.1200(h)(3)(iii)
(h)(3)(iii)
The measures employees can take to protect themselves from these hazards, including specific procedures the employer has implemented to protect employees from exposure to hazardous chemicals, such as appropriate work practices, emergency procedures, and personal protective equipment to be used; and,
(h)(3)(iv)
The details of the hazard communication program developed by the employer, including an explanation of the labeling system and the material safety data sheet, and how employees can obtain and use the appropriate hazard information.
(i)
"Trade secrets."
(i)(1)
The chemical manufacturer, importer, or employer may withhold the specific chemical identity, including the chemical name and other specific identification of a hazardous chemical, from the material safety data sheet, provided that:
(i)(1)(i)
The claim that the information withheld is a trade secret can be supported;
(i)(1)(ii)
Information contained in the material safety data sheet concerning the properties and effects of the hazardous chemical is disclosed;
(i)(1)(iii)
The material safety data sheet indicates that the specific chemical identity is being withheld as a trade secret; and,
(i)(1)(iv)
The specific chemical identity is made available to health professionals, employees, and designated representatives in accordance with the applicable provisions of this paragraph.

..1910.1200(i)(2)
(i)(2)
Where a treating physician or nurse determines that a medical emergency exists and the specific chemical identity of a hazardous chemical is necessary for emergency or first-aid treatment, the chemical manufacturer, importer, or employer shall immediately disclose the specific chemical identity of a trade secret chemical to that treating physician or nurse, regardless of the existence of a written statement of need or a confidentiality agreement. The chemical manufacturer, importer, or employer may require a written statement of need and confidentiality agreement, in accordance with the provisions of paragraphs (i)(3) and (4) of this section, as soon as circumstances permit.
(i)(3)
In non-emergency situations, a chemical manufacturer, importer, or employer shall, upon request, disclose a specific chemical identity, otherwise permitted to be withheld under paragraph (i)(1) of this section, to a health professional (i.e. physician, industrial hygienist, toxicologist, epidemiologist, or occupational health nurse) providing medical or other occupational health services to exposed employee(s), and to employees or designated representatives, if:
(i)(3)(i)
The request is in writing;
(i)(3)(ii)
The request describes with reasonable detail one or more of the following occupational health needs for the information:
(i)(3)(ii)(A)
To assess the hazards of the chemicals to which employees will be exposed;
(i)(3)(ii)(B)
To conduct or assess sampling of the workplace atmosphere to determine employee exposure levels;
(i)(3)(ii)(C)
To conduct pre-assignment or periodic medical surveillance of exposed employees;
(i)(3)(ii)(D)
To provide medical treatment to exposed employees;

..1910.1200(i)(3)(ii)(E)
(i)(3)(ii)(E)
To select or assess appropriate personal protective equipment for exposed employees;
(i)(3)(ii)(F)
To design or assess engineering controls or other protective measures for exposed employees; and,
(i)(3)(ii)(G)
To conduct studies to determine the health effects of exposure.
(i)(3)(iii)
The request explains in detail why the disclosure of the specific chemical identity is essential and that, in lieu thereof, the disclosure of the following information to the health professional, employee, or designated representative, would not satisfy the purposes described in paragraph (i)(3)(ii) of this section:
(i)(3)(iii)(A)
The properties and effects of the chemical;
(i)(3)(iii)(B)
Measures for controlling workers' exposure to the chemical;
(i)(3)(iii)(C)
Methods of monitoring and analyzing worker exposure to the chemical; and,

(i)(3)(iii)(D)
Methods of diagnosing and treating harmful exposures to the chemical;
(i)(3)(iv)
The request includes a description of the procedures to be used to maintain the confidentiality of the disclosed information; and,

..1910.1200(i)(3)(v)
(i)(3)(v)
The health professional, and the employer or contractor of the services of the health professional (i.e. downstream employer, labor organization, or individual employee), employee, or designated representative, agree in a written confidentiality agreement that the health professional, employee, or designated representative, will not use the trade secret information for any purpose other than the health need(s) asserted and agree not to release the information under any circumstances other than to OSHA, as provided in paragraph (i)(6) of this section, except as authorized by the terms of the agreement or by the chemical manufacturer, importer, or employer.
(i)(4)
The confidentiality agreement authorized by paragraph (i)(3)(iv) of this section:
(i)(4)(i)
May restrict the use of the information to the health purposes indicated in the written statement of need;
(i)(4)(ii)
May provide for appropriate legal remedies in the event of a breach of the agreement, including stipulation of a reasonable pre-estimate of likely damages; and,
(i)(4)(iii)
May not include requirements for the posting of a penalty bond.
(i)(5)
Nothing in this standard is meant to preclude the parties from pursuing non-contractual remedies to the extent permitted by law.
(i)(6)
If the health professional, employee, or designated representative receiving the trade secret information decides that there is a need to disclose it to OSHA, the chemical manufacturer, importer, or employer who provided the information shall be informed by the health professional, employee, or designated representative prior to, or at the same time as, such disclosure.

..1910.1200(i)(7)
(i)(7)
If the chemical manufacturer, importer, or employer denies a written request for disclosure of a specific chemical identity, the denial must:
(i)(7)(i)
Be provided to the health professional, employee, or designated representative, within thirty days of the request;
(i)(7)(ii)
Be in writing;
(i)(7)(iii)
Include evidence to support the claim that the specific chemical identity is a trade secret;
(i)(7)(iv)
State the specific reasons why the request is being denied; and,
(i)(7)(v)
Explain in detail how alternative information may satisfy the specific medical or occupational health need without revealing the specific chemical identity.
(i)(8)
The health professional, employee, or designated representative whose request for information is denied under paragraph (i)(3) of this section may refer the request and the written denial of the request to OSHA for consideration.
(i)(9)
When a health professional, employee, or designated representative refers the denial to OSHA under paragraph (i)(8) of this section, OSHA shall consider the evidence to determine if:

..1910.1200(i)(9)(i)
(i)(9)(i)
The chemical manufacturer, importer, or employer has supported the claim that the specific chemical identity is a trade secret;
(i)(9)(ii)
The health professional, employee, or designated representative has supported the claim that there is a medical or occupational health need for the information; and,
(i)(9)(iii)
The health professional, employee or designated representative has demonstrated adequate means to protect the confidentiality.
(i)(10)
(i)(10)(i)
If OSHA determines that the specific chemical identity requested under paragraph (i)(3) of this section is not a "bona fide" trade secret, or that it is a trade secret, but the requesting health professional, employee, or designated representative has a legitimate medical or occupational health need for the information, has executed a written confidentiality agreement, and has shown adequate means to protect the confidentiality of the information, the chemical manufacturer, importer, or employer will be subject to citation by OSHA.

..1910.1200(i)(10)(ii)
(i)(10)(ii)
If a chemical manufacturer, importer, or employer demonstrates to OSHA that the execution of a confidentiality agreement would not provide sufficient protection against the potential harm from the unauthorized disclosure of a trade secret specific chemical identity, the Assistant Secretary may issue such orders or impose such additional limitations or conditions upon the disclosure of the requested chemical information as may be appropriate to assure that the occupational health services are provided without an undue risk of harm to the chemical manufacturer, importer, or employer.
(i)(11)
If a citation for a failure to release specific chemical identity information is contested by the chemical manufacturer, importer, or employer, the matter will be adjudicated before the Occupational Safety and Health Review Commission in accordance with the Act's enforcement scheme and the applicable Commission rules of procedure. In accordance with the Commission rules, when a chemical manufacturer, importer, or employer continues to withhold the information during the contest, the Administrative Law Judge may review the citation and supporting documentation "in camera" or issue appropriate orders to protect the confidentiality of such matters.
(i)(12)
Notwithstanding the existence of a trade secret claim, a chemical manufacturer, importer, or employer shall, upon request, disclose to the Assistant Secretary any information which this section requires the chemical manufacturer, importer, or employer to make available. Where there is a trade secret claim, such claim shall be made no later than at the time the information is provided to the Assistant Secretary so that suitable determinations of trade secret status can be made and the necessary protections can be implemented.
(i)(13)
Nothing in this paragraph shall be construed as requiring the disclosure under any circumstances of process or percentage of mixture information which is a trade secret.

..1910.1200(j)
(j)
"Effective dates." Chemical manufacturers, importers, distributors, and employers shall be in compliance with all provisions of this section by March 11, 1994.

Note: The effective date of the clarification that the exemption of wood and wood products from the Hazard Communication standard in paragraph (b)(6)(iv) only applies to wood and wood products including lumber which will not be processed, where the manufacturer or importer can establish that the only hazard they pose to employees is the potential for flammability or combustibility, and that the exemption does not apply to wood or wood products which have been treated with a hazardous chemical covered by this standard, and wood which may be subsequently sawed or cut generating dust has been stayed from March 11, 1994 to August 11, 1994.

[59 FR 17479, April 13, 1994; 59 FR 65947, Dec. 22, 1994; 61 FR 5507, Feb. 13, 1996]

Appendix C

OSHA Regulations (Standards - 29 CFR) General Requirements. - 1910.132

Standard Number: 1910.132
Standard Title: General Requirements.
SubPart Number: I
SubPart Title: Personal Protective Equipment

(a)
Application. Protective equipment, including personal protective equipment for eyes, face, head, and extremities, protective clothing, respiratory devices, and protective shields and barriers, shall be provided, used, and maintained in a sanitary and reliable condition wherever it is necessary by reason of hazards of processes or environment, chemical hazards, radiological hazards, or mechanical irritants encountered in a manner capable of causing injury or impairment in the function of any part of the body through absorption, inhalation or physical contact.
(b)
Employee-owned equipment. Where employees provide their own protective equipment, the employer shall be responsible to assure its adequacy, including proper maintenance, and sanitation of such equipment.
(c)
Design. All personal protective equipment shall be of safe design and construction for the work to be performed.

..1910.132(d)
(d)
Hazard assessment and equipment selection.
(d)(1)
The employer shall assess the workplace to determine if hazards are present, or are likely to be present, which necessitate the use of personal protective equipment (PPE). If such hazards are present, or likely to be present, the employer shall:
(d)(1)(i)
Select, and have each affected employee use, the types of PPE that will protect the affected employee from the hazards identified in the hazard assessment;
(d)(1)(ii)
Communicate selection decisions to each affected employee; and,
(d)(1)(iii)
Select PPE that properly fits each affected employee. Note: Non-mandatory Appendix B contains an example of procedures that would comply with the requirement for a hazard assessment.
(d)(2)
The employer shall verify that the required workplace hazard assessment has been performed through a written certification that identifies the workplace evaluated; the person certifying that the evaluation has been performed; the date(s) of the hazard assessment; and, which identifies the document as a certification of hazard assessment.
(e)
Defective and damaged equipment. Defective or damaged personal protective equipment shall not be used.

..1910.132(f)
(f)
Training.
(f)(1)
The employer shall provide training to each employee who is required by this section to use PPE. Each such employee shall be trained to know at least the following:
(f)(1)(i)
When PPE is necessary;
(f)(1)(ii)
What PPE is necessary;
(f)(1)(iii)
How to properly don, doff, adjust, and wear PPE;
(f)(1)(iv)
The limitations of the PPE; and,
(f)(1)(v)
The proper care, maintenance, useful life and disposal of the PPE.
(f)(2)
Each affected employee shall demonstrate an understanding of the training specified in paragraph (f)(1) of this section, and the ability to use PPE properly, before being allowed to perform work requiring the use of PPE.
(f)(3)
When the employer has reason to believe that any affected employee who has already been trained does not have the understanding and skill required by paragraph (f)(2) of this section, the employer shall retrain each such employee. Circumstances where retraining is required include, but are not limited to, situations where:
(f)(3)(i)
Changes in the workplace render previous training obsolete; or

..1910.132(f)(3)(ii)
(f)(3)(ii)
Changes in the types of PPE to be used render previous training obsolete; or
(f)(3)(iii)
Inadequacies in an affected employee's knowledge or use of assigned PPE indicate that the employee has not retained the requisite understanding or skill.
(f)(4)
The employer shall verify that each affected employee has received and understood the required training through a written certification that contains the name of each employee trained, the date(s) of training, and that identifies the subject of the certification.
(g)
Paragraphs (d) and (f) of this section apply only to 1910.133, 1910.135, 1910.136, and 1910.138. Paragraphs (d) and (f) of this section do not apply to 1910.134 and 1910.137.

[59 FR 16360, April 6, 1994; 59 FR 33910, July 1, 1994; 59 FR 34580, July 6, 1994]

Appendix D

OSHA Regulations (Standards - 29 CFR) Eye and Face Protection. - 1910.133

Standard Number: 1910.133
Standard Title: Eye and Face Protection.
SubPart Number: I
SubPart Title: Personal Protective Equipment

(a)
General requirements.
(a)(1)
The employer shall ensure that each affected employee uses appropriate eye or face protection when exposed to eye or face hazards from flying particles, molten metal, liquid chemicals, acids or caustic liquids, chemical gases or vapors, or potentially injurious light radiation.
(a)(2)
The employer shall ensure that each affected employee uses eye protection that provides side protection when there is a hazard from flying objects. Detachable side protectors (e.g. clip-on or slide-on side shields) meeting the pertinent requirements of this section are acceptable.
(a)(3)
The employer shall ensure that each affected employee who wears prescription lenses while engaged in operations that involve eye hazards wears eye protection that incorporates the prescription in its design, or wears eye protection that can be worn over the prescription lenses without disturbing the proper position of the prescription lenses or the protective lenses.
(a)(4)
Eye and face PPE shall be distinctly marked to facilitate identification of the manufacturer.

..1910.133(a)(5)
(a)(5)
The employer shall ensure that each affected employee uses equipment with filter lenses that have a shade number appropriate for the work being performed for protection from injurious light radiation. The following is a listing of appropriate shade numbers for various operations.

Filter Lenses for Protection Against Radiant Energy

Operations	Electrode Size 1/32 in.	Arc Current	Minimum(*) Protective Shade
Shielded metal arc welding			
	Less than 3	Less than 60 ...	7
	3-5	60-160	8
	5-8	160-250	10
	More than 8	250-550	11
Gas metal arc welding and flux cored arc welding			
		less than 60 ...	7
		60-160	10
		160-250	10
		250-500	10
Gas Tungsten arc welding			
		less than 50 ...	8
		50-150	8
		150-500	10
Air carbon Arc cutting	(Light)	less than 500 ..	10
	(Heavy)	500-1000	11
Plasma arc welding			
		less than 20 ...	6
		20-100	8
		100-400	10
		400-800	11
Plasma arc cutting			
	(light)(**)	less than 300 ..	8
	(medium)(**)	300-400	9
	(heavy)(**)	400-800	10
Torch brazing		....	3
Torch soldering		....	2
Carbon arc welding		....	14

Filter Lenses for Protection Against Radiant Energy

Operations	Plate thickness-inches	Plate thickness-mm	Minimum(*) Protective Shade
Gas Welding:			
Light	Under 1/8	Under 3.2	4
Medium	1/8 to 1/2	3.2 to 12.7	5
Heavy	Over 1/2	Over 12.7	6
Oxygen cutting:			
Light	Under 1	Under 25	3
Medium	1 to 6	25 to 150	4
Heavy	Over 6	Over 150	5

Footnote(*) As a rule of thumb, start with a shade that is too dark to see the weld zone. Then go to a lighter shade which gives sufficient view of the weld zone without going below the minimum. In oxyfuel gas welding or cutting where the torch produces a high yellow light, it is desirable to use a filter lens that absorbs the yellow or sodium line in the visible light of the (spectrum) operation.

Footnote(**) These values apply where the actual arc is clearly seen. Experience has shown that lighter filters may be used when the arc is hidden by the workpiece.

(b)

Criteria for protective eye and face devices.

(b)(1)

Protective eye and face devices purchased after July 5, 1994 shall comply with ANSI Z87.1-1989, "American National Standard Practice for Occupational and Educational Eye and Face Protection," which is incorporated by reference as specified in Sec. 1910.6.

(b)(2)

Eye and face protective devices purchased before July 5, 1994 shall comply with the ANSI "USA standard for Occupational and Educational Eye and Face Protection," Z87.1-1968, which is incorporated by reference as specified in Sec. 1910.6, or shall be demonstrated by the employer to be equally effective.

[59 FR 16360, April 6, 1994; 59 FR 33910, July 1, 1994; 61 FR 9227, March 7, 1996; 61 FR 19547, May 2, 1996]

Appendix E

OSHA Regulations (Standards - 29 CFR) Respiratory Protection. - 1910.134

Standard Number: 1910.134
Standard Title: Respiratory Protection.
SubPart Number: I
SubPart Title: Personal Protective Equipment

This section applies to General Industry (part 1910), Shipyards (part 1915), Marine Terminals (part 1917), Longshoring (part 1918), and Construction (part 1926).

..1910.134(a)
(a)
Permissible practice.
(a)(1)
In the control of those occupational diseases caused by breathing air contaminated with harmful dusts, fogs, fumes, mists, gases, smokes, sprays, or vapors, the primary objective shall be to prevent atmospheric contamination. This shall be accomplished as far as feasible by accepted engineering control measures (for example, enclosure or confinement of the operation, general and local ventilation, and substitution of less toxic materials). When effective engineering controls are not feasible, or while they are being instituted, appropriate respirators shall be used pursuant to this section.
(a)(2)
Respirators shall be provided by the employer when such equipment is necessary to protect the health of the employee. The employer shall provide the respirators which are applicable and suitable for the purpose intended. The employer shall be responsible for the establishment and maintenance of a respiratory protection program which shall include the requirements outlined in paragraph (c) of this section.

..1910.134(b)
(b)
Definitions. The following definitions are important terms used in the respiratory protection standard in this section.
Air-purifying respirator means a respirator with an air-purifying filter, cartridge, or canister that removes specific air contaminants by passing ambient air through the air-purifying element.
Assigned protection factor (APF) [Reserved]
Atmosphere-supplying respirator means a respirator that supplies the respirator user with breathing air from a source independent of the ambient atmosphere, and includes supplied-air respirators (SARs) and self-contained breathing apparatus (SCBA) units.
Canister or cartridge means a container with a filter, sorbent, or catalyst, or combination of these items, which removes specific contaminants from the air passed through the container.
Demand respirator means an atmosphere-supplying respirator that admits breathing air to the facepiece only when a negative pressure is created inside the facepiece by inhalation.
Emergency situation means any occurrence such as, but not limited to, equipment failure, rupture of containers, or failure of control equipment that may or does result in an uncontrolled significant release of an airborne contaminant.
Employee exposure means exposure to a concentration of an airborne contaminant that would occur if the employee were not using respiratory protection.
End-of-service-life indicator (ESLI) means a system that warns the respirator user of the approach of the end of adequate respiratory protection, for example, that the sorbent is approaching saturation or is no longer effective.
Escape-only respirator means a respirator intended to be used only for emergency exit.
Filter or air purifying element means a component used in respirators to remove solid or liquid aerosols from the inspired air.
Filtering facepiece (dust mask) means a negative pressure particulate respirator with a filter as an integral part of the facepiece or with the entire facepiece composed of the filtering medium.
Fit factor means a quantitative estimate of the fit of a particular respirator to a specific individual, and typically estimates the ratio of the concentration of a substance in ambient air to its concentration inside the respirator when worn.
Fit test means the use of a protocol to qualitatively or quantitatively evaluate the fit of a respirator on an individual. (See also Qualitative fit test QLFT and Quantitative fit test QNFT.)
Helmet means a rigid respiratory inlet covering that also provides head protection against impact and penetration.
High efficiency particulate air (HEPA) filter means a filter that is at least 99.97% efficient in removing monodisperse particles of 0.3 micrometers in diameter. The equivalent NIOSH 42 CFR 84 particulate filters are the N100, R100, and P100 filters.
Hood means a respiratory inlet covering that completely covers the head and neck and may also cover portions of the shoulders and torso.
Immediately dangerous to life or health (IDLH) means an atmosphere that poses an immediate threat to life, would cause irreversible adverse health effects, or would impair an individual's ability to escape from a dangerous atmosphere.
Interior structural firefighting means the physical activity of fire suppression, rescue or both, inside of buildings or enclosed structures which are involved in a fire situation beyond the incipient stage. (See 29 CFR 1910.155)
Loose-fitting facepiece means a respiratory inlet covering that is designed to form a partial seal with the face. Maximum use concentration (MUC) [Reserved]. Negative pressure respirator (tight fitting) means a respirator in which the air pressure inside the facepiece is negative during inhalation with respect to the ambient air pressure outside the respirator.
Oxygen deficient atmosphere means an atmosphere with an oxygen content below 19.5% by volume.
Physician or other licensed health care professional (PLHCP) means an individual whose legally permitted scope of practice (i.e., license, registration, or certification) allows him or her to independently provide, or be delegated the responsibility to provide, some or all of the health care services required by paragraph (e) of this section.
Positive pressure respirator means a respirator in which the pressure inside the respiratory inlet covering exceeds the ambient air pressure outside the respirator.

Powered air-purifying respirator (PAPR) means an air-purifying respirator that uses a blower to force the ambient air through air-purifying elements to the inlet covering.
Pressure demand respirator means a positive pressure atmosphere-supplying respirator that admits breathing air to the facepiece when the positive pressure is reduced inside the facepiece by inhalation.
Qualitative fit test (QLFT) means a pass/fail fit test to assess the adequacy of respirator fit that relies on the individual's response to the test agent.
Quantitative fit test (QNFT) means an assessment of the adequacy of respirator fit by numerically measuring the amount of leakage into the respirator.
Respiratory inlet covering means that portion of a respirator that forms the protective barrier between the user's respiratory tract and an air-purifying device or breathing air source, or both. It may be a facepiece, helmet, hood, suit, or a mouthpiece respirator with nose clamp.
Self-contained breathing apparatus (SCBA) means an atmosphere-supplying respirator for which the breathing air source is designed to be carried by the user.
Service life means the period of time that a respirator, filter or sorbent, or other respiratory equipment provides adequate protection to the wearer.
Supplied-air respirator (SAR) or airline respirator means an atmosphere-supplying respirator for which the source of breathing air is not designed to be carried by the user.
This section means this respiratory protection standard.
Tight-fitting facepiece means a respiratory inlet covering that forms a complete seal with the face.
User seal check means an action conducted by the respirator user to determine if the respirator is properly seated to the face.

(c)

Respiratory protection program. This paragraph requires the employer to develop and implement a written respiratory protection program with required worksite-specific procedures and elements for required respirator use. The program must be administered by a suitably trained program administrator. In addition, certain program elements may be required for
voluntary use to prevent potential hazards associated with the use of the respirator. The Small Entity Compliance Guide contains criteria for the selection of a program administrator and a sample program that meets the requirements of this paragraph. Copies of the Small Entity Compliance Guide will be available on or about April 8, 1998 from the Occupational Safety and Health Administration's Office of Publications, Room N 3101, 200 Constitution Avenue, NW, Washington, DC, 20210 (202-219-4667).

(c)(1)

In any workplace where respirators are necessary to protect the health of the employee or whenever respirators are required by the employer, the employer shall establish and implement a written respiratory protection program with worksite-specific procedures. The program shall be updated as necessary to reflect those changes in workplace conditions that affect respirator use. The employer shall include in the program the following provisions of this section, as applicable:

(c)(1)(i)

Procedures for selecting respirators for use in the workplace;

(c)(1)(ii)

Medical evaluations of employees required to use respirators;

(c)(1)(iii)

Fit testing procedures for tight-fitting respirators;

(c)(1)(iv)

Procedures for proper use of respirators in routine and reasonably foreseeable emergency situations;

(c)(1)(v)

Procedures and schedules for cleaning, disinfecting, storing, inspecting, repairing, discarding, and otherwise maintaining
respirators;

(c)(1)(vi)

Procedures to ensure adequate air quality, quantity, and flow of breathing air for atmosphere-supplying respirators;

(c)(1)(vii)

Training of employees in the respiratory hazards to which they are potentially exposed during routine and emergency situations;

..1910.134(c)(1)(viii)

(c)(1)(viii)

Training of employees in the proper use of respirators, including putting on and removing them, any limitations on their use, and their maintenance; and

(c)(1)(ix)

Procedures for regularly evaluating the effectiveness of the program.

(c)(2)

Where respirator use is not required:

(c)(2)(i)

An employer may provide respirators at the request of employees or permit employees to use their own respirators, if the employer determines that such respirator use will not in itself create a hazard. If the employer determines that any voluntary respirator use is permissible, the employer shall provide the respirator users with the information contained in Appendix D to this section ("Information for Employees Using Respirators When Not Required Under the Standard"); and

(c)(2)(ii)

In addition, the employer must establish and implement those elements of a written respiratory protection program necessary to ensure that any employee using a respirator voluntarily is medically able to use that respirator, and that the respirator is cleaned, stored, and maintained so that its use does not present a health hazard to the user. Exception: Employers are not required to include in a written respiratory protection program those employees whose only use of respirators involves the voluntary use of filtering facepieces (dust masks).

(c)(3)

The employer shall designate a program administrator who is qualified by appropriate training or experience that is commensurate with the complexity of the program to administer or oversee the respiratory protection program and conduct the required evaluations of program effectiveness.

(c)(4)

The employer shall provide respirators, training, and medical evaluations at no cost to the employee.

(d)

Selection of respirators. This paragraph requires the employer to evaluate respiratory hazard(s) in the workplace, identify relevant workplace and user factors, and base respirator selection on these factors. The paragraph also specifies appropriately protective respirators for use in IDLH atmospheres, and limits the selection and use of air-purifying respirators.

(d)(1)

General requirements.

(d)(1)(i)

(i) The employer shall select and provide an appropriate respirator based on the respiratory hazard(s) to which the worker is exposed and workplace and user factors that affect respirator performance and reliability.

(d)(1)(ii)

The employer shall select a NIOSH-certified respirator. The respirator shall be used in compliance with the conditions of its certification.

(d)(1)(iii)

The employer shall identify and evaluate the respiratory hazard(s) in the workplace; this evaluation shall include a reasonable estimate of employee exposures to respiratory hazard(s) and an identification of the contaminant's chemical state and physical form. Where the employer cannot identify or reasonably estimate the employee exposure, the employer shall consider the atmosphere to be IDLH.

..1910.134(d)(1)(iv)

(d)(1)(iv)

The employer shall select respirators from a sufficient number of respirator models and sizes so that the respirator is acceptable to, and correctly fits, the user.

(d)(2)

Respirators for IDLH atmospheres.

(d)(2)(i)

The employer shall provide the following respirators for employee use in IDLH atmospheres:

(d)(2)(i)(A)

A full facepiece pressure demand SCBA certified by NIOSH for a minimum service life of thirty minutes, or

(d)(2)(i)(B)

A combination full facepiece pressure demand supplied-air respirator (SAR) with auxiliary self-contained air supply.

(d)(2)(ii)

Respirators provided only for escape from IDLH atmospheres shall be NIOSH-certified for escape from the atmosphere in which they will be

used.
(d)(2)(iii)
All oxygen-deficient atmospheres shall be considered IDLH. Exception: If the employer demonstrates that, under all foreseeable conditions, the oxygen concentration can be maintained within the ranges specified in Table II of this section (i.e., for the altitudes set out in the table), then any atmosphere-supplying respirator may be used.
(d)(3)
Respirators for atmospheres that are not IDLH.
(d)(3)(i)
The employer shall provide a respirator that is adequate to protect the health of the employee and ensure compliance with all other OSHA statutory and regulatory requirements, under routine and reasonably foreseeable emergency situations.
(d)(3)(i)(A)
Assigned Protection Factors (APFs) [Reserved]
(d)(3)(i)(B)
Maximum Use Concentration (MUC) [Reserved]
(d)(3)(ii)
The respirator selected shall be appropriate for the chemical state and physical form of the contaminant.
(d)(3)(iii)
For protection against gases and vapors, the employer shall provide:
(d)(3)(iii)(A)
An atmosphere-supplying respirator, or
(d)(3)(iii)(B)
An air-purifying respirator, provided that:
(d)(3)(iii)(B)(1)
(1) The respirator is equipped with an end-of-service-life indicator (ESLI) certified by NIOSH for the contaminant; or
(d)(3)(iii)(B)(2)
If there is no ESLI appropriate for conditions in the employer's workplace, the employer implements a change schedule for canisters and cartridges that is based on objective information or data that will ensure that canisters and cartridges are changed before the end of their service life. The employer shall describe in the respirator program the information and data relied upon and the basis for the canister and cartridge change schedule and the basis for reliance on the data.
(d)(3)(iv)
For protection against particulates, the employer shall provide:
(d)(3)(iv)(A)
An atmosphere-supplying respirator; or
(d)(3)(iv)(B)
An air-purifying respirator equipped with a filter certified by NIOSH under 30 CFR part 11 as a high efficiency particulate air (HEPA) filter, or an air purifying respirator equipped with a filter certified for particulates by NIOSH under 42 CFR part 84; or
(d)(3)(iv)(C)
For contaminants consisting primarily of particles with mass median aerodynamic diameters (MMAD) of at least 2 micrometers, an air-purifying respirator equipped with any filter certified for particulates by NIOSH.

TABLE I. -- Assigned Protection Factors [Reserved]

..1910.134(e)
(e)
Medical evaluation. Using a respirator may place a physiological burden on employees that varies with the type of respirator worn, the job and workplace conditions in which the respirator is used, and the medical status of the employee. Accordingly, this paragraph specifies the minimum requirements for medical evaluation that employers must implement to determine the employee's ability to use a respirator.
(e)(1)
General. The employer shall provide a medical evaluation to determine the employee's ability to use a respirator, before the employee is fit tested or required to use the respirator in the workplace. The employer may discontinue an employee's medical evaluations when the employee is no longer required to use a respirator.
(e)(2)
Medical evaluation procedures.
(e)(2)(i)
The employer shall identify a physician or other licensed health care professional (PLHCP) to perform medical evaluations using a medical questionnaire or an initial medical examination that obtains the same information as the medical questionnaire.
(e)(2)(ii)
The medical evaluation shall obtain the information requested by the questionnaire in Sections 1 and 2, Part A of Appendix C of this section.
(e)(3)
Follow-up medical examination.
(e)(3)(i)
The employer shall ensure that a follow-up medical examination is provided for an employee who gives a positive response to any question among questions 1 through 8 in Section 2, Part A of Appendix C or whose initial medical examination demonstrates the need for a follow-up medical examination.
(e)(3)(ii)
The follow-up medical examination shall include any medical tests, consultations, or diagnostic procedures that the PLHCP deems necessary to make a final determination.
(e)(4)
Administration of the medical questionnaire and examinations.
(e)(4)(i)
The medical questionnaire and examinations shall be administered confidentially during the employee's normal working hours or at a time and place convenient to the employee. The medical questionnaire shall be administered in a manner that ensures that the employee understands its content.
(e)(4)(ii)
The employer shall provide the employee with an opportunity to discuss the questionnaire and examination results with the PLHCP.
(e)(5)
Supplemental information for the PLHCP.
(e)(5)(i)
The following information must be provided to the PLHCP before the PLHCP makes a recommendation concerning an employee's ability to use a respirator:
(e)(5)(i)(A)
(A) The type and weight of the respirator to be used by the employee;
(e)(5)(i)(B)
The duration and frequency of respirator use (including use for rescue and escape);
(e)(5)(i)(C)
The expected physical work effort;
(e)(5)(i)(D)
Additional protective clothing and equipment to be worn; and
(e)(5)(i)(E)
Temperature and humidity extremes that may be encountered.
(e)(5)(ii)
Any supplemental information provided previously to the PLHCP regarding an employee need not be provided for a subsequent medical evaluation if the information and the PLHCP remain the same.
(e)(5)(iii)
The employer shall provide the PLHCP with a copy of the written respiratory protection program and a copy of this section.
Note to Paragraph (e)(5)(iii): When the employer replaces a PLHCP, the employer must ensure that the new PLHCP obtains this information, either by providing the documents directly to the PLHCP or having the documents transferred from the former PLHCP to the new PLHCP. However, OSHA does not expect employers to have employees medically reevaluated solely because a new PLHCP has been selected.
(e)(6)
Medical determination. In determining the employee's ability to use a respirator, the employer shall:
(e)(6)(i)
Obtain a written recommendation regarding the employee's ability to use the respirator from the PLHCP. The recommendation shall provide only the following information:
(e)(6)(i)(A)
Any limitations on respirator use related to the medical condition of the employee, or relating to the workplace conditions in which the respirator will be used, including whether or not the employee is medically able to use the respirator;
(e)(6)(i)(B)
The need, if any, for follow-up medical evaluations; and
(e)(6)(i)(C)
A statement that the PLHCP has provided the employee with a copy of the PLHCP's written recommendation.

(e)(6)(ii)
If the respirator is a negative pressure respirator and the PLHCP finds a medical condition that may place the employee's health at increased risk if the respirator is used, the employer shall provide a PAPR if the PLHCP's medical evaluation finds that the employee can use such a respirator; if a subsequent medical evaluation finds that the employee is medically able to use a negative pressure respirator, then the employer is no longer required to provide a PAPR.
(e)(7)
Additional medical evaluations. At a minimum, the employer shall provide additional medical evaluations that comply with the requirements of this section if:
(e)(7)(i)
An employee reports medical signs or symptoms that are related to ability to use a respirator;
(e)(7)(ii)
A PLHCP, supervisor, or the respirator program administrator informs the employer that an employee needs to be reevaluated;
(e)(7)(iii)
Information from the respiratory protection program, including observations made during fit testing and program evaluation, indicates a need for employee reevaluation; or
(e)(7)(iv)
A change occurs in workplace conditions (e.g., physical work effort, protective clothing, temperature) that may result in a substantial increase in the physiological burden placed on an employee.
(f)
Fit testing. This paragraph requires that, before an employee may be required to use any respirator with a negative or positive pressure tight-fitting facepiece, the employee must be fit tested with the same make, model, style, and size of respirator that will be used. This paragraph specifies the kinds of fit tests allowed, the procedures for conducting them, and how the results of the fit tests must be used.
(f)(1)
The employer shall ensure that employees using a tight-fitting facepiece respirator pass an appropriate qualitative fit test (QLFT) or quantitative fit test (QNFT) as stated in this paragraph.
(f)(2)
The employer shall ensure that an employee using a tight-fitting facepiece respirator is fit tested prior to initial use of the respirator, whenever a different respirator facepiece (size, style, model or make) is used, and at least annually thereafter.
(f)(3)
The employer shall conduct an additional fit test whenever the employee reports, or the employer, PLHCP, supervisor, or program administrator makes visual observations of, changes in the employee's physical condition that could affect respirator fit. Such conditions include, but are not limited to, facial scarring, dental changes, cosmetic surgery, or an obvious change in body weight.
(f)(4)
If after passing a QLFT or QNFT, the employee subsequently notifies the employer, program administrator, supervisor, or PLHCP that the fit of the respirator is unacceptable, the employee shall be given a reasonable opportunity to select a different respirator facepiece and to be retested.

..1910.134(f)(5)
(f)(5)
The fit test shall be administered using an OSHA-accepted QLFT or QNFT protocol. The OSHA-accepted QLFT and QNFT protocols and procedures are contained in Appendix A of this section.
(f)(6)
QLFT may only be used to fit test negative pressure air-purifying respirators that must achieve a fit factor of 100 or less.
(f)(7)
If the fit factor, as determined through an OSHA-accepted QNFT protocol, is equal to or greater than 100 for tight-fitting half facepieces, or equal to or greater than 500 for tight-fitting full facepieces, the QNFT has been passed with that respirator.
(f)(8)
Fit testing of tight-fitting atmosphere-supplying respirators and tight-fitting powered air-purifying respirators shall be accomplished by performing quantitative or qualitative fit testing in the negative pressure mode, regardless of the mode of operation (negative or positive pressure) that is used for respiratory protection.
(f)(1)(8)(i)
Qualitative fit testing of these respirators shall be accomplished by temporarily converting the respirator user's actual facepiece into a negative pressure respirator with appropriate filters, or by using an identical negative pressure air-purifying respirator facepiece with the same sealing surfaces as a surrogate for the atmosphere-supplying or powered air-purifying respirator facepiece.
(f)(1)(8)(ii)
Quantitative fit testing of these respirators shall be accomplished by modifying the facepiece to allow sampling inside the facepiece in the breathing zone of the user, midway between the nose and mouth. This requirement shall be accomplished by installing a permanent sampling probe onto a surrogate facepiece, or by using a sampling adapter designed to temporarily provide a means of sampling air from inside the facepiece.
(f)(1)(8)(iii)
Any modifications to the respirator facepiece for fit testing shall be completely removed, and the facepiece restored to NIOSH-approved configuration, before that facepiece can be used in the workplace.
(g)
Use of respirators. This paragraph requires employers to establish and implement procedures for the proper use of respirators. These requirements include prohibiting conditions that may result in facepiece seal leakage, preventing employees from removing respirators in hazardous environments, taking actions to ensure continued effective respirator operation throughout the work shift, and establishing procedures for the use of respirators in IDLH atmospheres or in interior structural firefighting situations.
(g)(1)
Facepiece seal protection.
(g)(1)(i)
The employer shall not permit respirators with tight-fitting facepieces to be worn by employees who have:
(g)(1)(i)(A)
Facial hair that comes between the sealing surface of the facepiece and the face or that interferes with valve function; or
(g)(1)(i)(B)
Any condition that interferes with the face-to-facepiece seal or valve function.
(g)(1)(ii)
If an employee wears corrective glasses or goggles or other personal protective equipment, the employer shall ensure that such equipment is worn in a manner that does not interfere with the seal of the facepiece to the face of the user.
(g)(1)(iii)
For all tight-fitting respirators, the employer shall ensure that employees perform a user seal check each time they put on the respirator using the procedures in Appendix B-1 or procedures recommended by the respirator manufacturer that the employer demonstrates are as effective as those in Appendix B-1 of this section.
(g)(2)
Continuing respirator effectiveness.
(g)(2)(i)
Appropriate surveillance shall be maintained of work area conditions and degree of employee exposure or stress. When there is a change in work area conditions or degree of employee exposure or stress that may affect respirator effectiveness, the employer shall reevaluate the continued effectiveness of the respirator.
(g)(2)(ii)
The employer shall ensure that employees leave the respirator use area:

..1910.134(g)(2)(ii)(A)
(g)(2)(ii)(A)
To wash their faces and respirator facepieces as necessary to prevent eye or skin irritation associated with respirator use; or
(g)(2)(ii)(B)
If they detect vapor or gas breakthrough, changes in breathing resistance, or leakage of the facepiece; or
(g)(2)(ii)(C)

To replace the respirator or the filter, cartridge, or canister elements.
(g)(2)(iii)
If the employee detects vapor or gas breakthrough, changes in breathing resistance, or leakage of the facepiece, the employer must replace or repair the respirator before allowing the employee to return to the work area.
(g)(3)
Procedures for IDLH atmospheres. For all IDLH atmospheres, the employer shall ensure that:
(g)(3)(i)
One employee or, when needed, more than one employee is located outside the IDLH atmosphere;
(g)(3)(ii)
Visual, voice, or signal line communication is maintained between the employee(s) in the IDLH atmosphere and the employee(s) located outside the IDLH atmosphere;
(g)(3)(iii)
The employee(s) located outside the IDLH atmosphere are trained and equipped to provide effective emergency rescue;
(g)(3)(iv)
The employer or designee is notified before the employee(s) located outside the IDLH atmosphere enter the IDLH atmosphere to provide emergency rescue;
(g)(3)(v)
The employer or designee authorized to do so by the employer, once notified, provides necessary assistance appropriate to the situation;
(g)(3)(vi)
Employee(s) located outside the IDLH atmospheres are equipped with:
(g)(3)(vi)(A)
Pressure demand or other positive pressure SCBAs, or a pressure demand or other positive pressure supplied-air respirator with auxiliary SCBA; and either
(g)(3)(vi)(B)
Appropriate retrieval equipment for removing the employee(s) who enter(s) these hazardous atmospheres where retrieval equipment would contribute to the rescue of the employee(s) and would not increase the overall risk resulting from entry; or
(g)(3)(vi)(C)
Equivalent means for rescue where retrieval equipment is not required under paragraph (g)(3)(vi)(B).
(g)(4)
Procedures for interior structural firefighting. In addition to the requirements set forth under paragraph (g)(3), in interior structural fires, the employer shall ensure that:
(g)(4)(I)
At least two employees enter the IDLH atmosphere and remain in visual or voice contact with one another at all times;
(g)(4)(ii)
At least two employees are located outside the IDLH atmosphere; and
(g)(4)(iii)
All employees engaged in interior structural firefighting use SCBAs.
Note 1 to paragraph (g): One of the two individuals located outside the IDLH atmosphere may be assigned to an additional role, such as incident commander in charge of the emergency or safety officer, so long as this individual is able to perform assistance or rescue activities without jeopardizing the safety or health of any firefighter working at the incident.
Note 2 to paragraph (g): Nothing in this section is meant to preclude firefighters from performing emergency rescue activities before an entire team has assembled.
(h)
(h) Maintenance and care of respirators. This paragraph requires the employer to provide for the cleaning and disinfecting, storage, inspection, and repair of respirators used by employees.
(h)(1)
Cleaning and disinfecting. The employer shall provide each respirator user with a respirator that is clean, sanitary, and in good working order. The employer shall ensure that respirators are cleaned and disinfected using the procedures in Appendix B-2 of this section, or procedures recommended by the respirator manufacturer, provided that such procedures are of equivalent effectiveness. The respirators shall be cleaned and disinfected at the following intervals:
(h)(1)(i)
Respirators issued for the exclusive use of an employee shall be cleaned and disinfected as often as necessary to be maintained in a sanitary condition;
(h)(1)(ii)
Respirators issued to more than one employee shall be cleaned and disinfected before being worn by different individuals;
(h)(1)(iii)
Respirators maintained for emergency use shall be cleaned and disinfected after each use; and
(h)(1)(iv)
Respirators used in fit testing and training shall be cleaned and disinfected after each use.
(h)(2)
Storage. The employer shall ensure that respirators are stored as follows:
(h)(2)(i)
All respirators shall be stored to protect them from damage, contamination, dust, sunlight, extreme temperatures, excessive moisture, and damaging chemicals, and they shall be packed or stored to prevent deformation of the facepiece and exhalation valve.
(h)(2)(ii)
In addition to the requirements of paragraph (h)(2)(i) of this section, emergency respirators shall be:
(h)(2)(ii)(A)
Kept accessible to the work area;
(h)(2)(ii)(B)
Stored in compartments or in covers that are clearly marked as containing emergency respirators; and
(h)(2)(ii)(C)
Stored in accordance with any applicable manufacturer instructions.

..1910.134(h)(3)
(h)(3)
Inspection.
(h)(3)(i)
The employer shall ensure that respirators are inspected as follows:
(h)(3)(i)(A)
All respirators used in routine situations shall be inspected before each use and during cleaning;
(h)(3)(i)(B)
All respirators maintained for use in emergency situations shall be inspected at least monthly and in accordance with the manufacturer's recommendations, and shall be checked for proper function before and after each use; and
(h)(3)(i)(C)
Emergency escape-only respirators shall be inspected before being carried into the workplace for use.
(h)(3)(ii)
The employer shall ensure that respirator inspections include the following:
(h)(3)(ii)(A)
A check of respirator function, tightness of connections, and the condition of the various parts including, but not limited to, the facepiece, head straps, valves, connecting tube, and cartridges, canisters or filters; and
(h)(3)(ii)(B)
A check of elastomeric parts for pliability and signs of deterioration.
(h)(3)(iii)
In addition to the requirements of paragraphs (h)(3)(i) and (ii) of this section, self-contained breathing apparatus shall be inspected monthly. Air and oxygen cylinders shall be maintained in a fully charged state and shall be recharged when the pressure falls to 90% of the manufacturer's recommended pressure level. The employer shall determine that the regulator and warning devices function properly.
(h)(3)(iv)
For respirators maintained for emergency use, the employer shall:
(h)(3)(iv)(A)
Certify the respirator by documenting the date the inspection was performed, the name (or signature) of the person who made the inspection, the findings, required remedial action, and a serial number or other means of identifying the inspected respirator; and (B) Provide this information on a tag or label that is attached to the storage compartment for the respirator, is kept with the respirator, or is included in inspection reports stored as paper or electronic files. This information shall be maintained until replaced following a subsequent certification.
(h)(4)
Repairs. The employer shall ensure that respirators that fail an inspec-

tion or are otherwise found to be defective are removed from service, and are discarded or repaired or adjusted in accordance with the following procedures:

(h)(4)(i)

Repairs or adjustments to respirators are to be made only by persons appropriately trained to perform such operations and shall use only the respirator manufacturer's NIOSH-approved parts designed for the respirator;

(h)(4)(ii)

Repairs shall be made according to the manufacturer's recommendations and specifications for the type and extent of repairs to be performed; and

(h)(4)(iii)

Reducing and admission valves, regulators, and alarms shall be adjusted or repaired only by the manufacturer or a technician trained by the manufacturer.

(i)

Breathing air quality and use. This paragraph requires the employer to provide employees using atmosphere-supplying respirators (supplied-air and SCBA) with breathing gases of high purity.

(i)(1)

The employer shall ensure that compressed air, compressed oxygen, liquid air, and liquid oxygen used for respiration accords with the following specifications:

(i)(1)(i)

Compressed and liquid oxygen shall meet the United States Pharmacopoeia requirements for medical or breathing oxygen; and

..1910.134(i)(1)(ii)

(i)(1)(ii)

Compressed breathing air shall meet at least the requirements for Grade D breathing air described in ANSI/Compressed Gas Association Commodity Specification for Air, G-7.1-1989, to include:

(i)(1)(ii)(A)

Oxygen content (v/v) of 19.5-23.5%;

(i)(1)(ii)(B)

Hydrocarbon (condensed) content of 5 milligrams per cubic meter of air or less;

(i)(1)(ii)(C)

Carbon monoxide (CO) content of 10 ppm or less;

(i)(1)(ii)(D)

Carbon dioxide content of 1,000 ppm or less; and

(i)(1)(ii)(E)

Lack of noticeable odor.

(i)(2)

The employer shall ensure that compressed oxygen is not used in atmosphere-supplying respirators that have previously used compressed air.

(i)(3)

The employer shall ensure that oxygen concentrations greater than 23.5% are used only in equipment designed for oxygen service or distribution.

(i)(4)

The employer shall ensure that cylinders used to supply breathing air to respirators meet the following requirements:

(i)(4)(i)

Cylinders are tested and maintained as prescribed in the Shipping Container Specification Regulations of the Department of Transportation (49 CFR part 173 and part 178);

(i)(4)(ii)

Cylinders of purchased breathing air have a certificate of analysis from the supplier that the breathing air meets the requirements for Grade D breathing air; and

(i)(4)(iii)

The moisture content in the cylinder does not exceed a dew point of -50 deg. F (-45.6 deg. C) at 1 atmosphere pressure.

(i)(5)

The employer shall ensure that compressors used to supply breathing air to respirators are constructed and situated so as to:

(i)(5)(i)

Prevent entry of contaminated air into the air-supply system;

(i)(5)(ii)

Minimize moisture content so that the dew point at 1 atmosphere pressure is 10 degrees F (5.56 deg.C) below the ambient temperature;

(i)(5)(iii)

Have suitable in-line air-purifying sorbent beds and filters to further ensure breathing air quality. Sorbent beds and filters shall be maintained and replaced or refurbished periodically following the manufacturer's instructions.

(i)(5)(iv)

Have a tag containing the most recent change date and the signature of the person authorized by the employer to perform the change. The tag shall be maintained at the compressor.

(i)(6)

For compressors that are not oil-lubricated, the employer shall ensure that carbon monoxide levels in the breathing air do not exceed 10 ppm.

(i)(7)

For oil-lubricated compressors, the employer shall use a high-temperature or carbon monoxide alarm, or both, to monitor carbon monoxide levels. If only high-temperature alarms are used, the air supply shall be monitored at intervals sufficient to prevent carbon monoxide in the breathing air from exceeding 10 ppm.

(i)(8)

The employer shall ensure that breathing air couplings are incompatible with outlets for nonrespirable worksite air or other gas systems. No asphyxiating substance shall be introduced into breathing air lines.

(i)(9)

The employer shall use breathing gas containers marked in accordance with the NIOSH respirator certification standard, 42 CFR part 84.

(j)

Identification of filters, cartridges, and canisters. The employer shall ensure that all filters, cartridges and canisters used in the workplace are labeled and color coded with the NIOSH approval label and that the label is not removed and remains legible.

(k)

Training and information. This paragraph requires the employer to provide effective training to employees who are required to use respirators. The training must be comprehensive, understandable, and recur annually, and more often if necessary. This paragraph also requires the employer to provide the basic information on respirators in Appendix D of this section to employees who wear respirators when not required by this section or by the employer to do so.

(k)(1)

The employer shall ensure that each employee can demonstrate knowledge of at least the following:

..1910.134(k)(1)(i)

(k)(1)(i)

Why the respirator is necessary and how improper fit, usage, or maintenance can compromise the protective effect of the respirator;

(k)(1)(ii)

What the limitations and capabilities of the respirator are;

(k)(1)(iii)

How to use the respirator effectively in emergency situations, including situations in which the respirator malfunctions;

(k)(1)(iv)

How to inspect, put on and remove, use, and check the seals of the respirator;

(k)(1)(v)

What the procedures are for maintenance and storage of the respirator;

(k)(1)(vi)

How to recognize medical signs and symptoms that may limit or prevent the effective use of respirators; and

(k)(1)(vii)

The general requirements of this section.

(k)(2)

The training shall be conducted in a manner that is understandable to the employee.

(k)(3)

The employer shall provide the training prior to requiring the employee to use a respirator in the workplace.

(k)(4)

An employer who is able to demonstrate that a new employee has received training within the last 12 months that addresses the elements specified in paragraph (k)(1)(i) through (vii) is not required to repeat such training provided that, as required by paragraph (k)(1), the employee can

demonstrate knowledge of those element(s). Previous training not repeated initially by the employer must be provided no later than 12 months from the date of the previous training.
(k)(5)
Retraining shall be administered annually, and when the following situations occur:
(k)(5)(i)
Changes in the workplace or the type of respirator render previous training obsolete;
(k)(5)(ii)
Inadequacies in the employee's knowledge or use of the respirator indicate that the employee has not retained the requisite understanding or skill; or
(k)(5)(iii)
Any other situation arises in which retraining appears necessary to ensure safe respirator use.
(k)(6)
The basic advisory information on respirators, as presented in Appendix D of this section, shall be provided by the employer in any written or oral format, to employees who wear respirators when such use is not required by this section or by the employer.
(l)
Program evaluation. This section requires the employer to conduct evaluations of the workplace to ensure that the written respiratory protection program is being properly implemented, and to consult employees to ensure that they are using the respirators properly.
(l)(1)
The employer shall conduct evaluations of the workplace as necessary to ensure that the provisions of the current written program are being effectively implemented and that it continues to be effective.
(l)(2)
The employer shall regularly consult employees required to use respirators to assess the employees' views on program effectiveness and to identify any problems. Any problems that are identified during this assessment shall be corrected. Factors to be assessed include, but are not limited to:

..1910.134((l)(2)(i)
(l)(2)(i)
Respirator fit (including the ability to use the respirator without interfering with effective workplace performance);
(l)(2)(ii)
Appropriate respirator selection for the hazards to which the employee is exposed;
(l)(2)(iii)
Proper respirator use under the workplace conditions the employee encounters; and
(l)(2)(iv)
Proper respirator maintenance.
(m)
Recordkeeping. This section requires the employer to establish and retain written information regarding medical evaluations, fit testing, and the respirator program. This information will facilitate employee involvement in the respirator program, assist the employer in auditing the adequacy of the program, and provide a record for compliance determinations by OSHA.

..1910.134(m)(1)
(m)(1)
Medical evaluation. Records of medical evaluations required by this section must be retained and made available in accordance with 29 CFR 1910.1020.
(m)(2)
Fit testing.
(m)(2)(i)
The employer shall establish a record of the qualitative and quantitative fit tests administered to an employee including:
(m)(2)(i)(A)
The name or identification of the employee tested;
(m)(2)(i)(B)
Type of fit test performed;
(m)(2)(i)(C)
Specific make, model, style, and size of respirator tested;
(m)(2)(i)(D)
Date of test; and
(m)(2)(i)(E)
The pass/fail results for QLFTs or the fit factor and strip chart recording or other recording of the test results for QNFTs.
(m)(2)(ii)
Fit test records shall be retained for respirator users until the next fit test is administered.
(m)(3)
A written copy of the current respirator program shall be retained by the employer.
(m)(4)
Written materials required to be retained under this paragraph shall be made available upon request to affected employees and to the Assistant Secretary or designee for examination and copying.
(n)
Dates.
(n)(1)
Effective date. This section is effective April 8, 1998. The obligations imposed by this section commence on the effective date unless otherwise noted in this paragraph. Compliance with obligations that do not commence on the effective date shall occur no later than the applicable start-up date.
(n)(2)
Compliance dates. All obligations of this section commence on the effective date except as follows:

..1910.134(n)(2)(i)
(n)(2)(i)
The determination that respirator use is required (paragraph (a)) shall be completed no later than September 8, 1998.
(n)(2)(ii)
Compliance with provisions of this section for all other provisions shall be completed no later than October 5, 1998.
(n)(3)
The provisions of 29 CFR 1910.134 and 29 CFR 1926.103, contained in the 29 CFR parts 1900 to 1910.99 and the 29 CFR part 1926 editions, revised as of July 1, 1997, are in effect and enforceable until October 5, 1998, or during any administrative or judicial stay of the provisions of this section.
(n)(4)
Existing Respiratory Protection Programs. If, in the 12 month period preceding April 8, 1998, the employer has conducted annual respirator training, fit testing, respirator program evaluation, or medical evaluations, the employer may use the results of those activities to comply with the corresponding provisions of this section, providing that these activities were conducted in a manner that meets the requirements of this section.

..1910.134(o)
(o)
Appendices.
(o)(1)
Compliance with Appendix A, Appendix B-1, Appendix B-2, and Appendix C of this section is mandatory.
(o)(2)
Appendix D of this section is non-mandatory and is not intended to create any additional obligations not otherwise imposed or to detract from any existing obligations.

[63 FR 1152, Jan. 8, 1998; 63 FR 20098, April 23, 1998]

Appendix F

OSHA Regulations (Standards - 29 CFR) Head Protection. - 1910.135

Standard Number: 1910.135
Standard Title: Head Protection.
SubPart Number: I
SubPart Title: Personal Protective Equipment

(a)
General requirements.
(a)(1)
The employer shall ensure that each affected employee wears a protective helmet when working in areas where there is a potential for injury to the head from falling objects.
(a)(2)
The employer shall ensure that a protective helmet designed to reduce electrical shock hazard is worn by each such affected employee when near exposed electrical conductors which could contact the head.
(b)
Criteria for protective helmets.
(b)(1)
Protective helmets purchased after July 5, 1994 shall comply with ANSI Z89.1-1986, "American National Standard for Personnel Protection-Protective Headwear for Industrial Workers-Requirements,"which is incorporated by reference as specified in Sec. 1910.6, or shall be demonstrated to be equally effective.
(b)(2)
Protective helmets purchased before July 5, 1994 shall comply with the ANSI standard "American National Standard Safety Requirements for Industrial Head Protection," ANSI Z89.1-1969, which is incorporated by reference as specified in Sec. 1910.6, or shall be demonstrated by the employer to be equally effective.

[59 FR 16362, April 6, 1994; 61 FR 9227, March 7, 1996; 61 FR 19547, May 1, 1996]

Appendix G

OSHA Regulations (Standards - 29 CFR) The Control of Hazardous Energy (lockout/tagout). - 1910.147

Standard Number: 1910.147
Standard Title:
The Control of Hazardous Energy (lockout/tagout).
SubPart Number: I
SubPart Title: Personal Protective Equipment

(a)
"Scope, application and purpose" -
(a)(1)
"Scope"
(a)(1)(i)
This standard covers the servicing and maintenance of machines and equipment in which the "unexpected" energization or start up of the machines or equipment, or release of stored energy could cause injury to employees. This standard establishes minimum performance requirements for the control of such hazardous energy.
(a)(1)(ii)
This standard does not cover the following:
(a)(1)(ii)(A)
Construction, agriculture and maritime employment;
(a)(1)(ii)(B)
Installations under the exclusive control of electric utilities for the purpose of power generation, transmission and distribution, including related equipment for communication or metering; and
(a)(1)(ii)(C)
Exposure to electrical hazards from work on, near, or with conductors or equipment in electric utilization installations, which is covered by Subpart S of this part; and

..1910.147(a)(1)(ii)(D)
(a)(1)(ii)(D)
Oil and gas well drilling and servicing.
(a)(2)
"Application."
(a)(2)(i)
This standard applies to the control of energy during servicing and/or maintenance of machines and equipment.
(a)(2)(ii)
Normal production operations are not covered by this standard (See Subpart O of this Part). Servicing and/or maintenance which takes place during normal production operations is covered by this standard only if:
(a)(2)(ii)(A)
An employee is required to remove or bypass a guard or other safety device; or
(a)(2)(ii)(B)
An employee is required to place any part of his or her body into an area on a machine or piece of equipment where work is
actually performed upon the material being processed (point of operation) or where an associated danger zone exists during a machine operating cycle.

Note: Exception to paragraph (a)(2)(ii): Minor tool changes and adjustments, and other minor servicing activities, which take place during normal production operations, are not covered by this standard if they are routine, repetitive, and integral to the use of the equipment for production, provided that the work is performed using alternative measures which provide effective protection (See Subpart O of this Part).

(a)(2)(iii)
This standard does not apply to the following:

..1910.147(a)(2)(iii)(A)
(a)(2)(iii)(A)
Work on cord and plug connected electric equipment for which exposure to the hazards of unexpected energization or start up of the equipment is controlled by the unplugging of the equipment from the energy source and by the plug being under the exclusive control of the employee performing the servicing or maintenance.
(a)(2)(iii)(B)
Hot tap operations involving transmission and distribution systems for substances such as gas, steam, water or petroleum products when they are performed on pressurized pipelines, provided that the employer demonstrates that-
(a)(2)(iii)(B)(1)
continuity of service is essential;
(a)(2)(iii)(B)(2)
shutdown of the system is impractical; and {3} documented procedures are followed, and special equipment is used which will provide proven effective protection for employees.
(a)(3)
"Purpose."
(a)(3)(i)
This section requires employers to establish a program and utilize procedures for affixing appropriate lockout devices or tagout devices to energy isolating devices, and to otherwise disable machines or equipment to prevent unexpected energization, start up or release of stored energy in order to prevent injury to employees.
(a)(3)(ii)
When other standards in this part require the use of lockout or tagout, they shall be used and supplemented by the procedural and training requirements of this section.
(b)
"Definitions applicable to this section."
"Affected employee." An employee whose job requires him/her to operate or use a machine or equipment on which servicing or maintenance is being performed under lockout or tagout, or whose job requires him/her

to work in an area in which such servicing or maintenance is being performed.
"Authorized employee." A person who locks out or tags out machines or equipment in order to perform servicing or maintenance on that machine or equipment. An affected employee becomes an authorized employee when that employee's duties include performing servicing or maintenance covered under this section.
"Capable of being locked out." An energy isolating device is capable of being locked out if it has a hasp or other means of attachment to which, or through which, a lock can be affixed, or it has a locking mechanism built into it. Other energy isolating devices are capable of being locked out, if lockout can be achieved without the need to dismantle, rebuild, or replace the energy isolating device or permanently alter its energy control capability.
"Energized." Connected to an energy source or containing residual or stored energy.
"Energy isolating device." A mechanical device that physically prevents the transmission or release or energy, including but not limited to the following: A manually operated electrical circuit breaker, a disconnect switch, a manually operated switch by which the conductors of a circuit can be disconnected from all ungrounded supply conductors and, in addition, no pole can be operated independently; a line valve; a block; and any similar device used to block or isolate energy. Push buttons, selector switches and other control circuit type devices are not energy isolating devices.
"Energy source." Any source of electrical, mechanical, hydraulic, pneumatic, chemical, thermal, or other energy.
"Hot tap." A procedure used in the repair maintenance and services activities which involves welding on a piece of equipment (pipelines, vessels or tanks) under pressure, in order to install connections or appurtenances. it is commonly used to replace or add sections of pipeline without the interruption of service for air, gas, water, steam, and petrochemical distribution systems.
"Lockout." The placement of a lockout device on an energy isolating device, in accordance with an established procedure, ensuring that the energy isolating device and the equipment being controlled cannot be operated until the lockout device is removed.
"Lockout device." A device that utilizes a positive means such as a lock, either key or combination type, to hold an energy isolating device in the safe position and prevent the energizing of a machine or equipment. Included are blank flanges and bolted slip blinds.
"Normal production operations." The utilization of a machine or equipment to perform its intended production function.
"Servicing and/or maintenance." Workplace activities such as constructing, installing, setting up, adjusting, inspecting, modifying, and maintaining and/or servicing machines or equipment. These activities include lubrication, cleaning or unjamming of machines or equipment and making adjustments or tool changes, where the employee may be exposed to the unexpected energization or startup of the equipment or release of hazardous energy.
"Setting up." Any work performed to prepare a machine or equipment to perform its normal production operation.
"Tagout." The placement of a tagout device on an energy isolating device, in accordance with an established procedure, to indicate that the energy isolating device and the equipment being controlled may not be operated until the tagout device is removed.
"Tagout device." A prominent warning device, such as a tag and a means of attachment, which can be securely fastened to an energy isolating device in accordance with an established procedure, to indicate that the energy isolating device and the equipment being controlled may not be operated until the tagout device is removed.

..1910.147(c)
(c)
"General" -
(c)(1)
"Energy control program." The employer shall establish a program consisting of energy control procedures, employee training and to periodic inspections to ensure that before any employee performs any servicing or maintenance on a machine or equipment where the unexpected energizing, startup or release of stored energy could occur and cause injury, the machine or equipment shall be isolated from the energy source and rendered inoperative.
(c)(2)
"Lockout/tagout."
(c)(2)(i)
If an energy isolating device is not capable of being locked out, the employer's energy control program under paragraph (c)(1) of this section shall utilize a tagout system.
(c)(2)(ii)
If an energy isolating device is capable of being locked out, the employer's energy control program under paragraph (c)(1) of this section shall utilize lockout, unless the employer can demonstrate that the utilization of a tagout system will provide full employee protection as set forth in paragraph (c)(3) of this section.
(c)(2)(iii)
After January 2, 1990, whenever replacement or major repair, renovation or modification of a machine or equipment is performed, and whenever new machines or equipment are installed, energy isolating devices for such machine or equipment shall be designed to accept a lockout device.
(c)(3)
"Full employee protection."
(c)(3)(i)
When a tagout device is used on an energy isolating device which is capable of being locked out, the tagout device shall be attached at the same location that the lockout device would have been attached, and the employer shall demonstrate that the tagout program will provide a level of safety equivalent to that obtained by using a lockout program.

..1910.147(c)(3)(ii)
(c)(3)(ii)
In demonstrating that a level of safety is achieved in the tagout program which is equivalent to the level of safety obtained by using a lockout program, the employer shall demonstrate full compliance with all tagout-related provisions of this standard together with such additional elements as are necessary to provide the equivalent safety available from the use of a lockout device. Additional means to be considered as part of the demonstration of full employee protection shall include the implementation of additional safety measures such as the removal of an isolating circuit element, blocking of a controlling switch, opening of an extra disconnecting device, or the removal of a valve handle to reduce the likelihood of inadvertent energization.
(c)(4)
"Energy control procedure."
(c)(4)(i)
Procedures shall be developed, documented and utilized for the control of potentially hazardous energy when employees are engaged in the activities covered by this section.

Note: "Exception:" The employer need not document the required procedure for a particular machine or equipment, when all of the following elements exist: [1] The machine or equipment has no potential for stored or residual energy or reaccumulation of stored energy after shut down which could endanger employees: [2] the machine or equipment has a single energy source which can be readily identified and isolated: [3] the isolation and locking out of that energy source will completely deenergize and deactivate the machine or equipment: [4] the machine or equipment is isolated from that energy source and locked out during servicing or maintenance: [5] a single lockout device will achieve a locker-out condition: [6] the lockout device is under the exclusive control of the authorized employee performing the servicing or maintenance: [7] the servicing or maintenance does not create hazards for other employees; and [8] the employer, in utilizing this exception, has had no accidents involving the unexpected activation or reenergization of the machine or equipment during servicing or maintenance.

(c)(4)(ii)
The procedures shall clearly and specifically outline the scope, purpose, authorization, rules, and techniques to be utilized for the control of hazardous energy, and the means to enforce compliance including, but not limited to, the following:
(c)(4)(ii)(A)
A specific statement of the intended use of the procedure;
(c)(4)(ii)(B)
Specific procedural steps for shutting down, isolating, blocking and securing machines or equipment to control hazardous energy;

(c)(4)(ii)(C)
Specific procedural steps for the placement, removal and transfer of lockout devices or tagout devices and the responsibility for them; and

..1910.147(c)(4)(II)(D)
(c)(4)(ii)(D)
Specific requirements for testing a machine or equipment to determine and verify the effectiveness of lockout devices, tagout devices, and other energy control measures.
(c)(5)
"Protective materials and hardware."
(c)(5)(i)
Locks, tags, chains, wedges, key blocks, adapter pins, self-locking fasteners, or other hardware shall be provided by the employer for isolating, securing or blocking of machines or equipment from energy sources.
(c)(5)(ii)
Lockout devices and tagout devices shall be singularly identified; shall be the only device(s) used for controlling energy; shall not be used for other purposes; and shall meet the following requirements:
(c)(5)(ii)(A)
"Durable."
(c)(5)(ii)(A)(1)
Lockout and tagout devices shall be capable of withstanding the environment to which they are exposed for the maximum period of time that exposure is expected.
(c)(5)(ii)(A)(2)
Tagout devices shall be constructed and printed so that exposure to weather conditions or wet and damp locations will not cause the tag to deteriorate or the message on the tag to become illegible.
(c)(5)(ii)(A)(3)
Tags shall not deteriorate when used in corrosive environments such as areas where acid and alkali chemicals are handled land stored.

..1910.147(c)(5)(ii)(B)
(c)(5)(ii)(B)
"Standardized." Lockout and tagout devices shall be standardized within the facility in at least one of the following criteria: Color; shape; or size; and additionally, in the case of tagout devices, print and format shall be standardized.
(c)(5)(ii)(C)
"Substantial" -
(c)(5)(ii)(C)(1)
"Lockout devices." Lockout devices shall be substantial enough to prevent removal without the use of excessive force or unusual techniques, such as with the use of bolt cutters or other metal cutting tools.
(c)(5)(ii)(C)(2)
"Tagout devices." Tagout devices, including their means of attachment, shall be substantial enough to prevent inadvertent or accidental removal. Tagout device attachment means shall be of a non-reusable type, attachable by hand, self-locking, and non-releasable with a minimum unlocking strength of no less than 50 pounds and having the general design and basic characteristics of being at least equivalent to a one-piece, all environment-tolerant nylon cable tie.
(c)(5)(ii)(D)
"Identifiable." Lockout devices and tagout devices shall indicate the identify of the employee applying the device(s).
(c)(5)(iii)
Tagout devices shall warn against hazardous conditions if the machine or equipment is energized and shall include a legend such as the following: "Do Not Start. Do Not Open. Do Not Close. Do Not Energize. Do Not Operate."

..1910.147(c)(6)
(c)(6)
"Periodic inspection."
(c)(6)(i)
The employer shall conduct a periodic inspection of the energy control procedure at least annually to ensure that the procedure and the requirements of this standard are being followed.
(c)(6)(i)(A)
The periodic inspection shall be performed by an authorized employee other than the one(s) utilizing the energy control procedure being inspected.
(c)(6)(i)(B)
The periodic inspection shall be conducted to correct any deviations or inadequacies identified.
(c)(6)(i)(C)
Where lockout is used for energy control, the periodic inspection shall include a review, between the inspector and each authorized employee, of that employee's responsibilities under the energy control procedure being inspected.
(c)(6)(i)(D)
Where tagout is used for energy control, the periodic inspection shall include a review, between the inspector and each authorized and affected employee, of that employee's responsibilities under the energy control procedure being inspected, and the elements set forth in paragraph (c)(7)(ii) of this section.

..1910.147(c)(6)(ii)
(c)(6)(ii)
The employer shall certify that the periodic inspections have been performed. The certification shall identify the machine or equipment on which the energy control procedure was being utilized, the date of the inspection, the employees included in the inspection, and the person performing the inspection.
(c)(7)
"Training and communication."
(c)(7)(i)
The employer shall provide training to ensure that the purpose and function of the energy control program are understood by employees and that the knowledge and skills required for the safe application, usage, and removal of the energy controls are acquired by employees. The training shall include the following:
(c)(7)(i)(A)
Each authorized employee shall receive training in the recognition of applicable hazardous energy sources, the type and magnitude of the energy available in the workplace, and the methods and means necessary for energy isolation and control.
(c)(7)(i)(B)
Each affected employee shall be instructed in the purpose and use of the energy control procedure.
(c)(7)(i)(C)
All other employees whose work operations are or may be in an area where energy control procedures may be utilized, shall be instructed about the procedure, and about the prohibition relating to attempts to restart or reenergize machines or equipment which are locked out or tagged out.
(c)(7)(ii)
When tagout systems are used, employees shall also be trained in the following limitations of tags:

..1910.147(c)(7)(ii)(A)
(c)(7)(ii)(A)
Tags are essentially warning devices affixed to energy isolating devices, and do not provide the physical restraint on those devices that is provided by a lock.
(c)(7)(ii)(B)
when a tag is attached to an energy isolating means, it is not to be removed without authorization of the authorized person responsible for it, and it is never to be bypassed, ignored, or otherwise defeated.
(c)(7)(ii)(C)
Tags must be legible and understandable by all authorized employees, affected employees, and all other employees whose work operations are or may be in the area, in order to be effective.
(c)(7)(ii)(D)
Tags and their means of attachment must be made of materials which will withstand the environmental conditions encountered in the workplace.
(c)(7)(ii)(E)
Tags may evoke a false sense of security, and their meaning needs to be understood as part of the overall energy control program.
(c)(7)(ii)(F)
Tags must be securely attached to energy isolating devices so that they cannot be inadvertently or accidentally detached during use.
(c)(7)(iii)
Employee retraining.

..1910.147(c)(7)(iii)(A)
(c)(7)(iii)(A)
Retraining shall be provided for all authorized and affected employees whenever there is a change in their job assignments, a change in machines, equipment or processes that present a new hazard, or when there is a change in the energy control procedures.
(c)(7)(iii)(B)
Additional retraining shall also be conducted whenever a periodic inspection under paragraph (c)(6) of this section reveals, or whenever the employer has reason to believe that there are deviations from or inadequacies in the employee's knowledge or use of the energy control procedures.
(c)(7)(iii)(C)
The retraining shall reestablish employee proficiency and introduce new or revised control methods and procedures, as necessary.
(c)(7)(iv)
The employer shall certify that employee training has been accomplished and is being kept up to date. The certification shall contain each employee's name and dates of training.
(c)(8)
"Energy isolation." Lockout or tagout shall be performed only by the authorized employees who are performing the servicing or maintenance.
(c)(9)
"Notification of employees." Affected employees shall be notified by the employer or authorized employee of the application and removal of lockout devices or tagout devices. Notification shall be given before the controls are applied, and after they are removed from the machine or equipment.

..1910.147(d)
(d)
"Application of control." The established procedures for the application of energy control (the lockout or tagout procedures) shall cover the following elements and actions and shall be done in the following sequence:
(d)(1)
"Preparation for shutdown." Before an authorized or affected employee turns off a machine or equipment, the authorized employee shall have knowledge of the type and magnitude of the energy, the hazards of the energy to be controlled, and the method or means to control the energy.
(d)(2)
"Machine or equipment shutdown." The machine or equipment shall be turned off or shut down using the procedures established for the machine or equipment. An orderly shutdown must be utilized to avoid any additional or increased hazard(s) to employees as a result of the equipment stoppage.
(d)(3)
"Machine or equipment isolation." All energy isolating devices that are needed to control the energy to the machine or equipment shall be physically located and operated in such a manner as to isolate the machine or equipment from the energy source(s).
(d)(4)
"Lockout or tagout device application."
(d)(4)(i)
Lockout or tagout devices shall be affixed to each energy isolating device by authorized employees.

..1910.147(d)(4)(ii)
(d)(4)(ii)
Lockout devices, where used, shall be affixed in a manner to that will hold the energy isolating devices in a "safe" or "off" position.
(d)(4)(iii)
Tagout devices, where used, shall be affixed in such a manner as will clearly indicate that the operation or movement of energy isolating devices from the "safe" or "off" position is prohibited.
(d)(4)(iii)(A)
Where tagout devices are used with energy isolating devices designed with the capability of being locked, the tag attachment shall be fastened at the same point at which the lock would have been attached.
(d)(4)(iii)(B)
Where a tag cannot be affixed directly to the energy isolating device, the tag shall be located as close as safely possible to the device, in a position that will be immediately obvious to anyone attempting to operate the device.

(d)(5)
"Stored energy."
(d)(5)(i)
Following the application of logout or tagout devices to energy isolating devices, all potentially hazardous stored or residual energy shall be relieved, disconnected, restrained, and otherwise rendered safe.

..1910.147(d)(5)(ii)
(d)(5)(ii)
If there is a possibility of reaccumulation of stored energy to a hazardous level, verification of isolation shall be continued until the servicing or maintenance is completed, or until the possibility of such accumulation no longer exists.
(d)(6)
"Verification of isolation." Prior to starting work on machines or equipment that have been locked out or tagged out, the authorized employee shall verify that isolation and deenergization of the machine or equipment have been accomplished.
(e)
"Release from lockout or tagout." Before lockout or tagout devices are removed and energy is restored to the machine or equipment, procedures shall be followed and actions taken by the authorized employee(s) to ensure the following:
(e)(1)
"The machine or equipment." The work area shall be inspected to ensure that nonessential items have been removed and to ensure that machine or equipment components are operationally intact.
(e)(2)
"Employees."
(e)(2)(i)
The work area shall be checked to ensure that all employees have been safely positioned or removed.
(e)(2)(ii)
Before lockout or tagout devices are removed and before machines or equipment are energized, affected employees shall be notified that the lockout or tagout devices have been removed.

..1910.147(e)(2)(iii)
(e)(2)(iii)
After lockout or tagout devices have been removed and before a machine or equipment is started, affected employees shall be notified that the lockout or tagout device(s) have been removed.
(e)(3)
"Lockout or tagout devices removal." Each lockout or tagout device shall be removed from each energy isolating device by the employee who applied the device. Exception to paragraph (e)(3). When the authorized employee who applied the lockout or tagout device is not available to remove it, that device may be removed under the direction of the employer, provided that specific procedures and training for such removal have been developed, documented and incorporated into the employer's energy control program. The employer shall demonstrate that the specific procedure shall include at least the following elements:
(e)(3)(i)
Verification by the employer that the authorized employee who applied the device is not at the facility:
(e)(3)(ii)
Making all reasonable efforts to contact the authorized employee to inform him/her that his/her lockout or tagout device has been removed; and
(e)(3)(iii)
Ensuring that the authorized employee has this knowledge before he/she resumes work at that facility.

..1910.147(f)
(f)
"Additional requirements."
(f)(1)
"Testing or positioning of machines, equipment or components thereof." In situations in which lockout or tagout devices must be temporarily removed from the energy isolating device and the machine or equipment energized to test or position the machine, equipment or component thereof, the following sequence of actions shall be followed:

(f)(1)(i)
Clear the machine or equipment of tools and materials in accordance with paragraph (e)(1) of this section;
(f)(1)(ii)
Remove employees from the machine or equipment area in accordance with paragraph (e)(2) of this section;
(f)(1)(iii)
Remove the lockout or tagout devices as specified in paragraph (e)(3) of this section;
(f)(1)(iv)
Energize and proceed with testing or positioning;
(f)(1)(v)
Deenergize all systems and reapply energy control measures in accordance with paragraph (d) of this section to continue the servicing and/or maintenance.
(f)(2)
"Outside personnel (contractors, etc.)."
(f)(2)(i)
Whenever outside servicing personnel are to be engaged in activities covered by the scope and application of this standard, the on-site employer and the outside employer shall inform each other of their respective lockout or tagout procedures.

..1910.147(f)(2)(ii)
(f)(2)(ii)
The on-site employer shall ensure that his/her employees understand and comply with the restrictions and prohibitions of the outside employer's energy control program.
(f)(3)
"Group lockout or tagout."
(f)(3)(i)
When servicing and/or maintenance is performed by a crew, craft, department or other group, they shall utilize a procedure which affords the employees a level of protection equivalent to that provided by the implementation of a personal lockout or tagout device.
(f)(3)(ii)
Group lockout or tagout devices shall be used in accordance with the procedures required by paragraph (c)(4) of this section including, but not necessarily limited to, the following specific requirements:
(f)(3)(ii)(A)
Primary responsibility is vested in an authorized employee for a set number of employees working under the protection of a group lockout or tagout device (such as an operations lock);
(f)(3)(ii)(B)
Provision for the authorized employee to ascertain the exposure status of individual group members with regard to the lockout or tagout of the machine or equipment and
(f)(3)(ii)(C)
When more than one crew, craft, department, etc. is involved, assignment of overall job-associated lockout or tagout control responsibility to an authorized employee designated to coordinate affected work forces and ensure continuity of protection; and

..1910.147(f)(3)(ii)(D)
(f)(3)(ii)(D)
Each authorized employee shall affix a personal lockout or tagout device to the group lockout device, group lockbox, or comparable mechanism when he or she begins work, and shall remove those devices when he or she stops working on the machine or equipment being serviced or maintained.
(f)(4)
"Shift or personnel changes." Specific procedures shall be utilized during shift or personnel changes to ensure the continuity of lockout or tagout protection, including provision for the orderly transfer of lockout or tagout device protection between off-going and oncoming employees, to minimize exposure to hazards from the unexpected energization or start-up of the machine or equipment, or the release of stored energy.

Note: The following Appendix to 1910.147 services as a non-mandatory guideline to assist employers and employees in complying with the requirements of this section, as well as to provide other helpful information. Nothing in the appendix adds to or detracts from any of the requirements of this section.

[61 FR 5507, Feb. 13, 1996]

Index

SAFETY NOTES

SAFETY NOTES

SAFETY NOTES